Mycorrhizal Biotechnology

T0225627

MYCORRHIZAL BIOTECHNOLOGY

Editors

Devarajan Thangadurai
Karnatak University, India

Carlos Alberto Busso
Universidad Nacional del Sur, Argentina

Mohamed Hijri
Université de Montréal, Canada

CRC Press
Taylor & Francis Group
Boca Raton London New York

CRC Press is an imprint of the
Taylor & Francis Group, an **informa** business

Science Publishers
Enfield, New Hampshire

CRC Press
Taylor & Francis Group
6000 Broken Sound Parkway NW, Suite 300
Boca Raton, FL 33487-2742

First issued in paperback 2017

ISBN-13: 978-1-57808-691-7 (hbk)
ISBN-13: 978-1-138-11281-0 (pbk)

Library of Congress Cataloging-in-Publication Data
Mycorrhizal biotechnology / editors, Devarajan Thangadurai,
Carlos Alberto Busso, Mohamed Hijri. -- 1st ed.
     p. cm.
  Includes bibliographical references and index.
  ISBN 978-1-57808-691-7 (hardcover)
  1. Mycorrhizas--Biotechnology. 2. Vesicular-arbuscular
  mycorrhizas--Biotechnology. I. Thangadurai, D.
  II. Busso, Carlos Alberto. III. Hijri, Mohamed.
  QK604.2.M962 2010
  660.6'2--dc22

Visit the Taylor & Francis Web site at
http://www.taylorandfrancis.com

and the CRC Press Web site at
http://www.crcpress.com

About the Editors

Devarajan Thangadurai is Assistant Professor of Botany at Karnatak University; President, Society for Applied Biotechnology; and Editor-in-Chief of *Biotechnology, Bioinformatics and Bioengineering*. He received his BSc in Biochemistry (Bharathiar University), MSc in Environmental Biotechnology (Manonmaniam Sundaranar University) and PhD in Plant Molecular Systematics (Sri Krishnadevaraya University) in India. From 1999 to 2004, he worked as a Junior and Senior Research Fellow with funding from the Ministry of Science and Technology, Government of India. Prior to joining the Karnatak University, he served as a Postdoctoral Fellow at the University of Madeira (Portugal), University of Delhi (India) and ICAR National Research Centre for Banana (India) from 2004 to 2006. He is the recipient of Acharya Nagarjuna University s 2003 Best Young Scientist Award with Gold Medal and Best Paper Presentation Award by the Bishop Hebbar College at Trichy in 2006. He has published over fifty research articles and authored fourteen books including Genetic Resources and Biotechnology (3 vols.); Genes, Genomes and Genomics (2 vols.); and Plant Stress and Biotechnology.

    Carlos Alberto Busso is Professor of Ecology at the Departamento de Agronomía-CERZOS (CONICET), Universidad Nacional del Sur (UNSur), Buenos Aires, Argentina. He is Editor-in-Chief of *Phyton - International Journal of Experimental Botany* and Associate Editor of the *Open Ecology Journal* and *Biodiversity Research International*. He received his Agronomy Engineer and MSc degrees in Plant Production at UNSur in 1978 and 1983 respectively, and his PhD degree in Range Ecology at Utah State University, USA. He started his work at UNSur as Teaching Assistant in Ecology in 1979, and continued with no interruption until today in the same discipline. He is currently teaching the undergraduate-level courses on General Ecology and Autoecology of Rangeland Plants, and the postgraduate level course on How to Write and Publish Scientific Articles. He conducted postdoctoral studies during 1995-96 at Texas A&M University, USA. He was a Faculty member of the Department of Agriculture, Biotechnology and Natural Resources at the University of Nevada during 2003-04 as a Research Assistant Professor. He has published more than eighty research articles (www.rangeecologybusso.com.ar) including books, over a wide spectrum of subjects in ecology.

Mohamed Hijri obtained his bachelor s degrees in Plant Biology (1993) and Cellular Biology (1994) at the Université Cadi Ayyadrank (Morocco) and Université de Bourgogne (France) respectively. He received his PhD in Cellular and Molecular Biology at the Université de Bourgogne in 1999. He worked as post-doctoral research fellow at the University of Basel and University of Lausanne in Switzerland from 2000-2004. Currently, he is Associate Professor of Biological Sciences at the Institute of Plant Biology, University of Montreal, Quebec, Canada and teaches Biodiversity, Mycology, Plant Pathology and Plant Protection. His research interests are in the areas of genetic structure, genome evolution, molecular identification and phylogeny of AM Fungi. He has authored over thirty research and review articles including two in the international weekly journal, *Nature*.

Preface

Mycorrhizae are the most common symbiotic species on earth in almost all ecosystems. They are symbioses between the roots of most plant species and fungi and today more than 90% of all plant taxa ranging from thallophyte to angiosperms form associations of one type or another with mycorrhizal fungi, as the productivity, species composition and the diversity of natural ecosystems are frequently dependent upon the presence and activity of mycorrhizae. Several studies have demonstrated that mycorrhizal associations play vital role in plant nutrition. They greatly increase the efficiency of nutrient and water uptake, enhance resistance to pathogens, buffer plant species against several environmental stresses and drought resistance. Mycorrhizae also improve plant growth and survival in soils contaminated by heavy metals.

These benefits of mycorrhizal symbioses, both agronomically by increased growth and yield as well as ecologically by improved fitness, indicate that mycorrhizal plants are often more competitive and better able to tolerate environmental stresses. While the term biotechnology has been used for the last three decades, major advances have generated new set of tools, which allow the development of commercial mycorrhizal inoculants for use in agriculture, horticulture, forestry and environmental reclamation due to a recent surge of interest in the subject by geneticists, biochemists, microbiologists and biotechnologists.

In this context, *Mycorrhizal Biotechnology* emphasizes strongly on the biotechnological application of mycorrhizae to promote the production of food while maintaining ecologically and economically sustainable production systems. It contains fourteen chapters written by leading experts in their respective fields of knowledge and expertise. In the first chapter, M. Miransari provides a review on the interactions of arbuscular mycorrhiza and soil microbes. In the second, L. Montecchio and L. Scattolin supply an overview on the ectomycorrhizae and forest plants fitness. In Chapter 3, V. Davamani, A.C. Lourduraj, R.P. Yogalakshmi and M. Velmurugan discuss the role of VAM in nutrient uptake of crop plants. In the next chapter A. Khaliq, D.J. Bagyaraj and M. Alam elaborate the mass production technology of arbuscular mycorrhiza. Biomass production, arbuscular mycorrhizae and soil plant-available P under water stress in perennial grasses is the subject matter of Chapter 5 by C.A. Busso and A.I. Bolletta. Y. Ismail and M. Hijri discuss about the role of arbuscular mycorrhizal fungi on induced resistance in plants in Chapter 6. Influence of VAM in bioremediation of environmental pollutants is the theme of seventh chapter by V. Davamani, A.C. Lourduraj and M. Velmurugan.

Rhizosphere management practices in sustaining mycorrhizae are discussed by C. Harisudan, M. Velmurugan and P. Hemalatha (Chapter 8). The next chapter deals with importance of mycorrhizae for horticultural crops by P. Hemalatha, M. Velmurugan, C. Harisudan and V. Davamani. In Chapter 10, M.S. Velázquez and M.N. Cabello discuss about mycobization. Biotechnological procedures involving plants and arbuscular mycorrhizal fungi are explained in chapter 11 (S.E. Hassan, M. St-Arnaud, M. Labreque and M. Hijri). Molecular characterization of genetic diversity among arbuscular mycorrhizal fungi is explained in the next chapter by M. Velmurugan, P. Hemalatha, C. Harisudan and D. Thangadurai. In Chapter 13, the use of primers to detect genetic diversity in arbuscular mycorrhizal fungi is discussed by F. Covacevich. Finally, the impact of climate change on arbuscular mycorrhiza is detailed in the last chapter by F. Covacevich and R.L.L. Berbara.

The task of editing and publishing this book has received the generous technical, scientific and editorial support of many individuals who provided extremely valuable assistance which is highly appreciated and gratefully acknowledged. The editors wish to thank many individuals for their contribution of time and expertise to the completion of this book. Special thanks are also to Dr. F. Covacevich for providing photographs to the cover page and Professor T. Pullaiah for his continuous encouragement and insightful thoughts.

Devarajan Thangadurai
Carlos Alberto Busso
Mohamed Hijri

Contents

1

Arbuscular Mycorrhiza and Soil Microbes

Mohammad Miransari

Department of Soil Science, College of Agricultural Sciences
Shahed University, Tehran, Iran

Introduction

Soil rhizosphere is a very interesting and complicated environment surrounding plant roots. There are very many different types of microorganisms in the soil rhizosphere interacting with the other microbes and with plant roots. The properties of soil rhizosphere make it a unique and active area. The activity and interactions of rhizotrophic microorganisms can very much influence soil conditions and hence plant growth and microorganism activities [1].

Arbuscular mycorrhizas (AM) are among the most important and influencing soil microbes significantly affecting the growth of plants and other soil microorganisms. The soil part around the plant roots and AM hypha, where AM and bacteria are interactive is called 'mycorrhizosphere'. There are also different types of soil bacteria in the soil, which are interactive with AM, particularly in the rhizosphere and in most cases the interactions are synergistic.

Consideration of the rhizotrophic interactions and their consequent effects on the soil properties and hence plant growth can have very important implications in agriculture and ecology. There are different effects resulted by such interactions, which modified soil structural properties [2] and soil enhanced availability of nutrients [3] are among the most important ones. Thus, it is pertinent to evaluate such interactions precisely and suggest some new perspectives for the future research, which can make the advance of the field more rapidly and result in more efficient agricultural strategies.

Arbuscular Mycorrhiza

Arbuscular mycorrhiza (AM) are zygomycetes belonging to the order Glomales. According to both fossil discoveries and DNA sequences, the appearance of both AM and plants is almost 400 million years old [4]. AM are able to develop symbiotic association with most terrestrial plants [5-8] and usually their symbiosis with the host plant is not host specific. In their symbiosis, the host plants provide the fungi with their required hydrocarbon and receive nutrients, especially P from the fungi [6, 9, 10].

The passage of AM into the plant roots is through their hypha, which would eventually form the arbuscules and the vesicles. Arbuscules are branched structural hypha, which are the place of nutrient exchange with the plant roots. Vesicles are the specialized storage organelles with numerous and large vacuoles, which can greatly help the host plant especially under different stresses such as salinity and heavy metals. The onset of the symbiosis and the beneficial effects of the two symbionts become likely through the communication of some signal molecules [9, 11].

AM are able to enhance plant tolerance to different stresses such as soil salinity and drought, soil compaction, heavy metals and pathogens [12-19]. Plant responses to different species of AM are different, and since AM greatly affect the variousness, biomass and nutrients uptake of plants, AM species determine very much the structure of plant communities [20, 21]. Different efficiency of plants species in symbiosis with AM species affects their ecological functioning [5, 12, 18-22].

AM is able to greatly influence plants structure, hence combination recognition of AM species, in symbiosis with plant species is very important. Although the AM-plant symbiosis is not specific, the probability of some symbiotic combinations is more likely. Compared with other AM-plant symbiosis some combinations are more common under field conditions [21].

Arbuscular Mycorrhiza and Soil Bacteria

Although AM are very important symbionts to plants and their symbiosis can significantly enhance the growth of the host plant, they are also very much interactive with different bacteria including both the bacterial strains in the rhizosphere and in the bacterial strains in the cytoplasm of some fungal species [23]. Understanding such interactions, particularly in agricultural producing areas with not much resources of input is of great significance and is a very interesting research topic, and in the meanwhile complicated, in the field of molecular microbe-plant interactions [23]. Relative to the complete recognition of symbiotic stages, activated during the process of N fixation between rhizobium and the specific host plant, including the exchange of the signal molecules between the two partners [24, 25], there are much more details, which should yet be elucidated regarding the symbiosis of AM with the host plant [23].

However, it should be mentioned that some very interesting achievements regarding the establishment of symbiosis between AM and the host plant including the recognition of the signal molecules involved, have been recently achieved [26]. In addition, the recognition of genes such as phosphate transporter genes [27] and metallothionein producing genes [28] related to the functional properties of AM are of great significance.

The obstacles related to mycorrhizal research including their mandatory biotrophic nature, their cellular structure and their variable genetical properties have made the advance of the field less rapid, relative to the rhizobium-legume symbiosis [23, 29]. In addition to the large size of AM genome (0.3 pg to 1.12 pg/DNA) that has made the performance of AM genomic projects unlikely [23]; the significance of the host plant should also be noticed. It is because the first plant that has been genetically sequenced is *Arabidopsis thaliana*, which is not a host to AM [30]. However, plants including rice are host to AM and, hence, the research related to rice genomic data can illustrate some very important knowledge regarding AM symbiosis [31].

Different researches have indicated that there are about 100,000 genes related to AM plant host symbiosis. Additionally, using mutants have indicated the presence of proteins, which are communicated between the two partners at the time of symbiosis establishment [32]. Also the discovery of genes such as NORK and SYMRK have made the identification of pathways related to signal perception by the host plant and bacteria and AM likely [23].

Arbuscular mycorrhiza are also great niches for other soil microbes and while some of the bacteria are attached to AM hypha, some of them are bound to plant roots [33, 34]. While for other eukaryotic cells the association with bacteria is very common [35], for AM fungi only a few strains of bacteria are integrated into the fungi [36, 37]. There are bacterial-like structures in the cytoplasm of AM [38, 39]. The presence of bacteria in AM species taken from the field has been indicated by ultra structural ways.

It is very important to determine the bacterial population, associated with AM fungi, with the highest physiological activities. This indicates the bacterial strains that are more efficient, particularly when interactive with AM, and can make the use of effective co-inoculation likely. There are different methods for the determination of bacterial association with AM. For example, for tagging and visualizing the bacterial strain *Paenibacillus brasilensis*, which has suppressing effects on the activity of plant pathogens and can stimulate the activity of some specific AM species, von der Weid et al. [40] used the green fluorescence protein technique.

Arbuscular mycorrhiza and bacterial interactions can take place in the rhizosphere before the onset of inoculation or after the establishment of the tripartite symbiosis between AM, bacteria and the host plant [39]. The synergistic interactions of AM and bacteria can stimulate plant growth through enhancing processes such as nutrients uptake and controlling plant pathogens. These processes are of great significance, especially in agricultural cropping strategies, which are not depending much on agrochemicals to maintain soil fertility and health. In addition, AM are also able to influence the combinations of soil bacterial populations [41]. These effects can be related to the alteration of root physiology by affecting the chemical combination of root products [42, 43].

Additionally, AM can also alter the combination of bacteria in the rhizosphere through competition for soil nutrients [44]. Scientists have also stated that the association of some bacteria with AM is specific [41] indicating that there are some kind of communications between the bacteria and AM, stimulated by fungus exudates [39]. This is also verified by the results of scientists who found that some bacterial genera including *Arthrobacter* and *Bacillus* were most common in the hyphosphere or the soil around specific AM hypha, while *Pseudomonas* spp. were most distributed in the *Sorghum bicolour* rhizosphere. This suggests that the likelihood of Gram-positive bacterial association with AM is higher related to the Gram-negative bacteria, but has yet to be verified [41].

Additionally, it has been stated that bacterial genera are more frequent in the rhizosphere than hyphosphere indicating that root exudates can be more beneficial to the bacteria than hyphal products [39]. There are different examples of enhancing association between bacteria and AM including species of *Bacillus*, *Paenibacillus*, *Pseudomonas* and *Rhizobia* that are in association with AM species of Glomus including *G. calrum*, *G. intraradices*, *G. mosseae*, *G. versiforme*, and *G. intraradices*. These stimulating effects include the growth and germination of fungi and spores, respectively, root colonization of the host plant by AM, the solubilization of phosphate, and the suppression of pathogens [39].

The significance of bacterial attachment to the AM hypha and whether it can affect hyphal growth has yet to be elucidated. However, if this is the case, the co-inoculation of appropriate bacteria with AM can significantly contribute to enhanced plant growth [39]. In addition to the nutritional effects of fungal products on bacterial growth, production of glycoproteins such as glomalin can also influence bacterial growth in the soil through improving soil structure [2]. For example, the bacteria, present in the water soluble aggregates are different from the bacteria, present in the non-soil water aggregates [39]. Bacterial types, which are interactive with AM, are saprophytes and symbionts that some of which are unfavourable, some are neutral and some are favourable [45].

The bacteria that are able to enhance plant growth through interacting with plant roots are called plant growth promoting rhizobacteria (PGPR). Although it has been indicated that some of the PGPR are able to perfectly inoculate plant roots, data related to the inoculation intensity of AM hypha by PGPR is little. According to Bianciotto et al. [46] the intensity attachment of some species of *Rhizobium* and *Pseudomonas* to AM germinating spore and hypha under sterilized conditions was different depending on the strains of bacteria. However, the level of specificity was not recognized.

The strength of bacterial binding to AM hypha differs during the different physiological stages of attachment including a weak electrostatic attachment in the first stage followed by a strong attachment in the second stage, which is related to the production of cellulose or other extracellular products by bacteria. This hypothesis is supported by the less strong attachment of bacterial mutants, which were not able to produce such products, to the AM hypha [39]. Because some bacterial strains such as *Pseudomonas* spp. are able to colonize both plant roots and AM hypha, it has been suggested that the related processes can be relatively similar.

Although, there has been a lot of research on the interactive activities between AM and bacteria, more research should be conducted to more clearly elucidate the processes involved in the interactions between AM and soil bacteria. This can be very useful for the optimum determination of bio-inoculants, necessary for sustainable agricultural cropping strategies [39]. In addition to the stimulating effects of PGPR on plant growth through enhancing nutrients uptake, suppressing pathogens and some beneficial biochemicals, their interactive effects with AM can also increase plant growth [39]. Researchers have stated that PGPR can have some very stimulatory effects on AM growth [47]. This indicates that the co-inoculation of AM and specific PGPR can enhance the activity of AM during the symbiosis with the host plant [39]. This is because for example, some PGPR such as *Pseudomonas putida* are able to stimulate the root inoculation of host plants by AM [48].

The interaction between AM and bacteria, particularly PGPR [49, 50], and N-fixing bacteria are very beneficial to the host plants. The data regarding the simultaneous and enhancing effects of AM and PGPR on plant growth is little. It has just been recently that these effects have been tested, simultaneously [39]. Some of the bacteria are able to influence AM activity of spore germination and growth rate [51, 52] and hence affecting plant growth through AM symbiosis.

Arbuscular Mycorrhiza and N-fixing Bacteria

Unlike AM, which develop symbiosis with most terrestrial plants, N-fixing bacteria are able to establish symbiotic association with their specific host plant and fix atmospheric N [4, 21, 24, 25, 53]. The N-fixing bacterial symbionts including the genera *Azorhizobium, Bradyrhizobium, Mezorhizobium, Rhizobium* and *Sinorhizobium*, collectively called rhizobia, are settled in a plant membrane including compartment called symbiosome [4, 54]. Similar to AM-plant symbiosis the bacteria-legume symbiosis is also of very important agricultural and environmental implications, because it can substantially contribute to N production and utilization [55, 56].

Process of N Fixation

Through some very interesting and in the meanwhile complicated biochemical dialogue, the bacteria realize the presence of the specific host plant roots and chemotactically approach the plant roots. The stages related to the development of symbiosis between the N-fixing bacteria and the host plant include: the root exudation of signal molecules, the gene activation of bacteria by the exudates [24, 25] and production of some biochemicals by the bacteria [57] that can induce morphogenesis and physiological changes in the host roots. This eventually results in the formation of nodules, which are the place of bacterial settlement and hence N fixation [24, 25].

Although N fixation by rhizobial symbiosis supplies the host plant with additional N, high amounts of energy and P must be available to the symbiosis process [21]. Different species of AM are able to provide the host plant with various amounts of P [58-60]. Arbuscular mycorrhiza are also able to mineralize organic N in the mycorrhizosphere and increase N availability to the plant [39].

It is believed that the N-fixing capability of *Rhizobium* may enhance if the host plant is also in symbiosis with AM. Under such a situation and with regard to enhancing the colonization rate, uptake of inorganic nutrients and plant growth, *Rhizobium* and AM are synergistic [39]. There are some common genetical stages for both bacterial N fixation and AM symbiosis. The extent of plant accommodation for the intracellular settlement of endosymbionts including AM and bacteria has yet to be recognized [4]. Accordingly, there are some differences between the bacteria and AM for the preparation of the symbiosis with their partner. For example, the weakening strength of the root cell walls for the passage of the bacteria into the roots is not common for AM passage into the roots [4].

Legumes may prefer to develop symbiosis with AM species, which are more efficient to supply P. This can be very advantageous under the conditions that nutrients are not available at high amounts [61]. Different AM species are able to increase nodulation and N fixation differently [21, 62]. The beneficial effect of AM on P uptake by *Rhizobium* has been proved through providing the non-mycorrhizal plants with additional P resulting in enhanced plant growth, which is comparable with the growth of mycorrhizal plants [39, 60].

The researches regarding AM variousness in legumes and their colonization of legume nodules are not much. Scientists have indicated that AM are able to colonize root nodules in the laboratory [63]. However, nodules colonization of AM under field conditions and the species of AM, which are able to colonize legumes nodules, is yet to be recognized. The structure, functioning, and nutritional demand of nodules are different with plant roots. Nodules are produced by cortical cell division, in which rhizobia with high energy and P requirements reside and fix N [21].

Nodule formation also alters plant physiological properties and induces plant systematic acquired resistance [21, 64] suggesting that AM species combination may be different in roots and nodules. AM communities in legume roots are different with non-legume roots. Some AM species are able to develop symbiosis with plants containing high amounts of N. This is also in agreement with the finding that N fertilization can alter the combination of AM species and also with the finding that the level of *G. intraradices* with the Glo8 sequence type increased, after fertilization with N and P fertilizers [65].

While N-fixing bacteria utilize dicarboxylates as their source of energy [66], AM absorb [67] glucose and fructose, just when residing in plant roots, and hence hexoses are translocated to the AM fungi [10]. Endosymbiosis is accompanied with the production of a high amount of symbiosome membrane, for example the area of plasmamembrane for a nodule cell with bacteroids is about 2800 μM^2, however the area for the symbiosome membrane is 21,500 μM^2 [68]. Additionally, since different nutrients must be translocated into the cells containing the symbionts, different compositions of proteins including different transporters are located within the symbiosome membrane.

Also the prebacteroid [69] and the prearbuscular membranes have ATPase activities, which are not available in pathogenic interactions [70]. Thus, the symbiosome membrane form two different components, one for the inclusion of the symbionts and one for the control of different proteins compositions and their activities. While the mutants of *Bradyrhizobium japonicum* are not able to activate under low oxygen conditions, the bacteria is able to develop a mechanism by which they can be active and fix N [71] under such conditions. Different mutants are categorized based on lacking the ability to synthesize cytochromes (necessary for bacteroids formation), pass through the infection thread, and transcribe different plant genes necessary for nodulins formation [72].

The symbiosis of AM species with legume plants make them more efficient, for example through enhancing nutrient uptake such as P, Cu and Zn, which are very important for nodulation and N fixation. Hence, the specificity between AM, and its host plant can be very important, as legumes would develop symbiosis with AM species, which are more efficient [21]. The tripartite symbiosis between AM, bacteria and legumes is of great significance both for agriculture and for ecology, and scientists have been trying to find the most efficient combination of AM and bacteria [21].

Although the colonization of legume nodules by AM have been proved in the laboratory, there is very little relevant data under field conditions. Compared with plant roots, the combination of AM species was unique for legume nodules; however the combination was similar for three different legume species. Although sequence type Glo8 was found more in legume nodules, sequence type Acau5 was just found in legume nodules. It is also worth mentioning that legume roots exclusively contained sequence type 50. These all suggest that legume nodules may have enhancing or inhibiting effects on root colonization of some AM species [21].

The following reasons indicate the specificity of legume nodules in their association with AM species: (1) the high tendency of nodules for nutrients such as P, Cu or Zn, might result in the association of some AM species with legume nodules, (2) high nodule N concentration may be preferable to some AM species (for example AM species with Glo8 and Acau5 sequences), (3) the similarities between the stages of nodule and AM symbiosis, as there are some very important common stages [21, 22] and hence, legume mutants, which are not able to develop symbioses with rhizobia, may not also develop symbioses with AM species [73], (4) AM symbiosis development is influenced by rhizobial signals [74], and (5) the different physiology of nodules and roots and also the alteration of root exudates by rhizobial symbiosis and the induction of systematic acquired resistance in plants. These are all the likely explanations indicating the different tendency of root nodules for different AM species as all affect the ability of AM species to colonize legume nodules [21]. They also found that AM species isolated from the field were able to colonize legume nodules by forming hypha around the nodules and producing hypha and spores in the nodules.

Plant host species determine the combination of AM species in legume roots indicating that the tendency of different host plants for symbiosis development with different AM species differs. Some AM species (with Glo8 and Glo3 sequences) have the ability to colonize a wide range of host legume plants in both roots and nodules [75]. Using the same PCR methods the variousness of AM species ranged from 0.4 to 2.3, using the Shannon index under cultivated field and tropical rain forest conditions [76]. The parameter most influencing the Shannon index is the specific plant species rather than the number of plant species. Hence, parameters such as the nature of plants functionality (legumes or non-legumes), species of plants and root components including roots and nodules can determine AM communities in plants. More research is required to specify the effects of different AM species combinations on plant and nodule performance, and also on the structure and combination of plants [21].

Seed inoculation of legumes with N-fixing bacteria including *rhizobia* and *bradyrhizobium* can be of some very practical applications as the method has turned into a very useful applicable technology in both developed and developing countries [77]. The environmental and economical approaches of N fixation are also very significant, as for example by application of less N fertilizer and the beneficial interacting effects of bacteria with other soil microorganisms including AM.

As previously mentioned usually legume plants are able to make tripartite symbiosis with N-fixing bacteria and AM. This kind of association can be very beneficial both to the plant and to the symbiotic microbes. To make the tripartite symbiosis highly efficient the interactions between the plants, the bacteria and the AM must exactly be elucidated. Scientists [78, 79] have indicated some interesting aspects related to such kind of symbiosis.

However, almost few researches have indicated the beneficial effects of AM, both local and inoculated on bacteria-legumes symbiosis. AM are able to transfer the absorbed N from the soil to the plant. It has been stated that AM are able to mineralize organic N present in organic matter and hence make soil N more available to the bacteria interactive with AM. However, the organic matter utilizing and mineralizing by AM and hence plant and AM nutrient uptake, which is related to the stimulating effects of AM on mineralizing bacteria, has yet to be elucidated [80].

The network of mycorrhizal hypha are able to transfer N between and within the plants. In addition to the important role of AM in cycling the nutrients, their contribution to intensive agriculture should also be clearly indicated [7].

By enhancing N uptake and hence plant growth and also through affecting bacterial dependence on atmospheric N, AM may affect N fixation. Under non-sterile conditions, legume roots can develop symbiosis with AM, which can help the plants such as forage and crop legumes absorb higher rate of N. AM and rhizobium can synergistically and significantly affect the symbiotic related parameters [7]. The different responses of legume species to AM inoculation have also been attributed to different root morphology and architecture and also different dependency on AM [81]. Compared with control, AM inoculation, plant dry matter, fixed N, P and K uptake were significantly increased in different varieties of *Phaseolus vulgaris* and *Vicia faba*, inoculated with a mixture of *R. leguminosarum bv. phaseoli* and *R. tropici* and also mixed species of *Glomus clarum*, *G. etunicatum*, *G. manihotis* and *Gigaspora margarita* [82]. Since P uptake is one of the nutrients, most affected by AM symbiosis, in mycorrhizal-legume plants the enhanced P concentration, especially during the seedling and reproductive stages can be very beneficial to the host plant. AM symbiosis is most effective in soils where the amount of available P is at low or medium levels.

The data regarding the effect of AM on micronutrients uptake when in a tripartite symbiosis with rhizobium and legumes is very few. However, AM and rhizobium inoculation increased the uptake of Zn and Cu, hyphal colonization, nodule dry weight, N uptake, fixed N and P uptake [78]. Most researches regarding the effects of AM symbiosis on plant growth have been conducted under controlled and greenhouse conditions, and there have been few researches regarding the effects of AM symbiosis on plant growth under field conditions. Hence, it is very necessary to develop non-mycorrhizal legumes mutants so that the effects of inoculated AM in the presence of local AM on legumes plants can also be exactly elucidated under field conditions. The mutants that have been developed so far have not been very applicable, because the mutated gene in the mutants (MYC$^-$) controls both the AM symbiosis and the nodule formation (nod$^-$) [7].

The other very important point is the recognition role of signal molecules such as flavonoid biochemicals that can have on AM and bacteria and legumes symbiosis [24, 25, 78, 79, 83, 84]. For example, the signal molecule formononetin is commercially available and has been indicated to be very effective on AM symbiosis [7, 84]. There are common legume plant genes and biochemical molecules affecting the tripartite symbiosis of the legume host plant with AM and N-fixing bacteria. The presence of a tripartite symbiosis in legumes indicates that legumes have some kind of genetical controlling processes that make this kind of symbiosis likely [85]. There are some pea mutants (*Pisum sativum* L.) that are not able to develop symbiosis with both AM and N-fixing bacteria. There are stages in both symbioses, controlled by the same genes including hyphal passage into the plant roots by appressorium (Myc1), the development of arbuscule and AM development rate [85, 86].

Effects of Bacteria on P Utilization by AM

Plant P transporters are located in the periarbuscular membrane. Scientists have already cloned the cDNAs from the roots of *M. truncatula*, which are able to activate P transporters. However, such cDNAs are not expressed in mycorrhizal roots indicating the P transporters that they are active in P uptake under non-symbiotic conditions only [87].

There are some bacteria, soil P solubilizing bacteria, in the soil that are able to enhance P uptake by AM and plant through enhancing the solubility of soil P, present in organic and inorganic matter. Organic and inorganic P is made available by phosphatase and organic acid producing bacteria, respectively. This significantly increases P uptake by AM hypha and hence the symbiotic host plant [60]. Soil inorganic form of P, which is not available to plants, is strongly bound in the insoluble structures of P and also is attached to the clay surface layers.

The synergistic effects of AM and soil solubilizing P bacteria has been indicated by different researches [88]. Under limited availability of soil P, the interaction effects between P solubilizing bacteria and AM result in the enhanced plant colonization by the host plant and the increased bacterial population in the rhizosphere. The coinoculation of AM and P solubilizing bacteria increased plant N and P uptake, relative to control plants [39].

Symbiosis and Signaling

The symbiosis of AM and N-fixing bacteria with the host plants are the most important mutual symbioses—agriculturally and ecologically [54]. For the onset of symbiosis between the host plants and microbes the exchange of signal molecules is necessary [24, 25]. Signal molecules are biochemical compounds stimulating the activity of different genes, involved in the process of symbiosis. Symbiotic bacteria or AM spores realize the presence of the host plant in the soil through the secretion of these molecules by the plant roots. Accordingly, the symbiotic bacteria make a chemotactic move to the roots and the spores begin to germinate. However, the spores may also germinate in the absence of the host plant, but they are not able to proceed with the following stages of symbiosis. The stimulated genes of symbiotic bacteria (nodulation genes, *NOD*, regulated by the transcriptional factor NodD) produce some signal molecules, called lipo-chito-oligosaccharides, which alter the cell cycling in the cortex of plant roots cells and hence resulting in morphological and physiological changes and eventually nodule formation [24, 25, 57].

Although because of the differences in host specificity between bacteria and AM the responses of the two symbionts are different, scientists have found that there are many common stages between bacterial and AM symbioses with the host plant including the exchange of signal molecules [54]. For example, as previously mentioned, there are identical genes in both symbioses that are activated during the symbiosis [89]. In addition, some of the signal molecules necessary for the AM-plant symbiosis have been recently identified [26]. There should be more than a single gene for the perception of Nod (bacteria) and Myc (AM) factors. Because, rhizobium and AM result in different responses, including morphological and although both factors induce similar genes, Nod factors require some extra genes for inducing responses in the host plant. Accordingly, Nod and Myc factors induce genes in different parts of the plant root including epidermis and cortex, respectively, indicating that different plant receptors perceive Nod and Myc factors. Thus the two factors must result in the activation of different pathways in the plant [54]. Since legumes are able to develop symbiotic association with both N-fixing bacteria and AM, they are ideal for the study of the common stages of symbiosis between the two symbionts [90].

Conclusion

Accordingly, the following conclusions may be drawn. The great importance of the interactions among the host plant AM and soil bacteria taking place in soil is more indicated. These interactions must be clearly elucidated as they can have some very significant implications in agriculture and ecology. In addition to their individual functioning in the soil the combined effects of soil microbes are also very important as for example for the production of bio-inoculants or their enhancing effects on soil structure and plant nutrients uptake can increase plant growth and hence crop yield. So future research may focus more precisely on the interactions between the host plant, AM and soil bacteria for the more efficient use of soil microorganisms for the development of very advanced agricultural strategies.

References

1. Zaidi, A., Khan, M.S. and Amil, M.D. *European Journal of Agronomy* 2003, 19:15-21.
2. Rillig, M.C. and Mummey, D.L. *New Phytologist* 2006, 171:41-53.
3. Marschener, H. and Dell, B. *Plant and Soil* 1994, 159:89-102.
4. Parniske, M. *Current Opinion in Plant Biology* 2000, 3:320-328.
5. van der Heijden, M.G.A., Klironomos, J.N., Ursic, M. et al. *Nature* 1998, 396:69-72.
6. Hause, B., Maier, W., Miersch, O. et al. *Plant Physiology* 2002, 130:1213-1220.
7. Chalk, P.M., Souza, R., Urquiaga, S. et al. *Soil Biology and Biochemistry* 2006, 38:2944-2951.
8. Pongrac, P., Vogel-Mikus, K., Kump, P. et al. *Chemosphere* 2007, 69:1602-1609.
9. Harrison, M.J. *Annual Reviews in Physiology and Plant Molecular Biology* 1999, 50:361-389.
10. Harrison, M.J. *Journal of Experimental Botany* 1999, 50:1013-1022.
11. Matusova, R., Rani, K., Verstappen, F.W.A. et al. *Plant Physiology* 2005, 139:920-934.
12. Davies, F.T., Potter, J.R. and Linderman, R.G. *Physiolgia Plantarum* 1993, 87:45-53.
13. Auge, R.M. *Mycorrhiza* 2001, 11:3-42.
14. Feng, G., Zhang, F.S., Li, X.L. et al. *Mycorrhiza* 2002, 12:185-190.
15. Citterio, S., Prato, N., Fumagalli, P. et al. *Chemosphere* 2005, 59:21-29.
16. Subramanian, K., Santhanakrishnan, P. and Balasubramanian, P. *Scientia Horticulturae* 2006, 107:245-253.
17. Hildebrandt, U., Regvar, M. and Bothe, H. *Phytochemistry* 2007, 68:139-146.
18. Miransari, M., Bahrami, H.A., Rejali, F. et al. *Soil Biology and Biochemistry* 2007, 39:2014-2026.
19. Miransari, M., Bahrami, H.A., Rejali, F. et al. *Soil Biology and Biochemistry*, 2008, 40:1197-1206.
20. van der Heijden, M.G.A., Boller, T., Wiemken, A. et al. *Ecology* 1998, 79:2082-2091.
21. Scheublin, T.R., Ridgway, K.P., Young, P.W. et al. *Applied and Environmental Microbiology* 2004, 70:6240-6246.
22. Provorov, N.A., Borisov, A.Y. and Tikhonovich, I.A. *Journal of Theoretical Biology* 2002, 214:215-232.
23. Bonfante, P. *Biological Bulletin* 2003, 204:215-220.
24. Miransari, M. and Smith, D.L. *Journal of Plant Nutrition* 2007, 30:1967-1992.
25. Miransari, M. and Smith, D.L. *Journal of Plant Interactions*, 2008, 3:287-295.
26. Akiyama, K. and Hayashi, H. *Annals of Botany* 2006, 97:925-931.
27. Harrison, M.J. and van Buuren, M.L. *Nature* 1995, 378:26-29.
28. Lanfranco, L.A., Bolchi, E., Cesale Ros, S. et al. *Plant Physiology* 2002, 130:58-67.
29. Hijri, M., Hosny, M., van Tuinen D. et al. *Fungal Genetical Biology* 1999, 26:141-151.
30. Arabidopsis Genome Initiative. *Nature* 2000, 408:796-815.
31. Goodman, R.M., Naylor, R., Tefera, H. et al. *Science* 2002, 296:1801-1804.

32. Stracke, S., Kistner, C., Yoshida, S. et al. *Nature* 2002, 417:959-962.
33. Bianciotto, V., Andreotti, S., Balestrini, R. et al. *Molecular Plant-Microbe Interactions* 2001, 14:255-260.
34. Bianciotto, V., Lumini, E., Bonfante, P. et al. *International Journal of Systematic and Evolutionary Microbiology* 2003, 53:121-126.
35. Moran, N.A. and Wernegreen, J.J. *Tree* 2000, 15:321-326.
36. Schüssler, A. and Kluge, M. *In:* The Mycota IX. Hock, B. (ed.), Springer Verlag, Berlin, 2001, pp. 151-161.
37. de Boer, W., Folman, L.B., Summerbell, R.C. et al. *FEMS Microbiology Reviews* 2005, 29:795-811.
38. Mosse, B. *Archives of Microbiology* 1970, 74:129-145.
39. Artursson, V., Finlay, R.D. and Jansson, J.K. *Environmental Microbiology* 2006, 8:1-10.
40. Von der Weid, I., Artursson, V., Seldin, L. et al. *World Journal of Microbiological Biotechnology* 2005, 21:1591-1597.
41. Artursson, V., Finlay, R.D. and Jansson, J.K. *Environmental Microbiology* 2005, 7:1952-1966.
42. Gryndler, M. *In:* Arbuscular Mycorrhizas: Physiology and Function. Kapulnik, Y. and Douds, D.D.J. (eds). Kluwer Academic Publishers, Dordrecht, the Netherlands, 2000, pp. 239-262.
43. Linderman, R.G. *In:* Arbuscular Mycorrhizas: Physiology and Function, Kapulnik, Y. and Douds, D.D.J. (eds.), Kluwer Academic Publishers, Dordrecht, the Netherlands, 2000, pp. 345-365.
44. Christensen, H. and Jakobsen, I. *Biology and Fertility of Soils* 1993, 15:253-258.
45. Johansson, J.F., Paul, L.R. and Finlay, R.D. *FEMS Microbiological Ecology* 2004, 48:1-13.
46. Bianciotto, V., Minerdi, D., Perotto, S. et al. *Protoplasma* 1996, 193:123-131.
47. Linderman, R.G. *In:* The Mycota. Caroll, G.C. and Tudzynski, P. (eds), Springer-Verlag, Berlin, Germany, 1997, pp. 117-128.
48. Meyer, J.R. and Linderman, R.G. *Soil Biology and Biochemistry* 1986, 18:185-190.
49. von Alten, H., Lindermann, A. and Schonbeck, F. *Mycorrhiza* 1993, 2:167-173.
50. Kloepper, J.W. *Bioscience* 1996, 46:406-409.
51. Mosse, B. *Transactions in British Mycological Society* 1959, 42:273-286.
52. Carpenter-Boggs, L., Loynachan, T.E. and Stahl, P.D. *Soil Biology and Biochemistry* 1995, 27:1445-1451.
53. Sprent, J.I. Nodulation in legumes. Royal Botanical Gardens, Kew, 2001.
54. Limpens, E. and Bisseling, T. *Current Opinion in Plant Biology* 2003, 6:343-350.
55. Vandermeer, J.H. The ecology of intercropping. Cambridge University Press, New York, 1989.
56. Cleveland, C.C., Townsend, A.R., Schimel, D.S. et al. *Global Biogeochemistry Cycles* 1999, 13:623-645.
57. Miransari, M., Smith, D.L., Mackenzie, A.F. et al. *Communications in Soil Science and Plant Analysis* 2006, 37:1103-1110.
58. Jakobsen, I., Abbott, L.K. and Robson, A.D. *New Phytologist* 1992, 120:371-380.
59. Ravnskov, S. and Jakobsen, I. *New Phytologist* 1995, 129:611-618.

60. Smith, S.E. and Read, D.J. Mycorrhizal Symbiosis, 2nd ed., Academic Press, London, 1997.
61. Schulze, J. *Journal of Plant Nutrition and Soil Science* 2004, 167:125-137.
62. Ianson, D.C. and Linderman, R.G. *Symbiosis* 1993, 15:105-119.
63. Baird, L.M. and Caruso, K.J. *International Journal of Plant Sciences* 1994, 155:633-639.
64. Lian, B., Zhou, X. and Miransari, M. *Journal of Agronomy and Crop Science* 2000, 185:187-192.
65. Johnson, N.C. *Ecological Applications* 1993, 3:749-757.
66. Udvardi, M.K. and Day, D.A. *Annual Reviews in Plant Physiology and Plant Molecular Biology* 1997, 48:493-523.
67. Pfeffer, P.E., Douds, D.D., Becard, G. et al. *Plant Physiology* 1999, 120:587-598.
68. Roth, L.E. and Stacey, G. *European Journal of Cell Biology* 1989, 49:13-23.
69. Fedorova, E., Thomson, R., Whitehead, L.F. et al. *Planta* 1999, 209:25-32.
70. Whitehead, L.F. and Day, D.A. *Physiologia Plantarum* 1997, 100:30-44.
71. Preisig, O., Zufferey, R., Thöny-Meyer, L. et al. *Journal of Bacteriology* 1996, 178:1532-1538.
72. Ramseier, T.M., Winteler, H.V. and Hennecke, H. *Journal of Biological Chemistry* 1991, 266:7793-7803.
73. Duc, G., Trouvelot, A., Gianinazzi-Pearson, V. et al. *Plant Science* 1989, 60:215-222.
74. Xie, Z., Staehelin, C., Vierheilig, H. et al. *Plant Physiology* 1995, 108:1519-1525.
75. Opik, M., Moora, M., Liira, J. et al. *New Phytologist* 2003, 160:581-593.
76. Husband, R., Herre, E.A., Turner, S.L. et al. *Molecular Ecology* 2002, 11:2669-2678.
77. Alves, B.J.R., Boddey, R.M. and Urquiaga, S. *Plant and Soil* 2003, 252:1-9.
78. Antunes, P.M., de Varennes, A., Rajcan, I. et al. *Soil Biology and Biochemistry* 2006, 38:1234-1242.
79. Antunes, P.M., Rajcan, I. and Goss, M.J. *Soil Biology and Biochemistry* 2006, 38:533-543.
80. Hodge, A., Campbell, C.D. and Fitter, A.H. *Nature* 2001, 413:297-299.
81. Schoeneberger, M.M., Volk, R.J. and Davey, C.B. *Soil Science Society of America Journal* 1989, 53:1429-1434.
82. Ibijbijen, J., Urquiaga, S., Ismaili, M. et al. *New Phytologist* 1996, 134:353-360.
83. Zhang, F. and Smith, D.L. *Plant Physiology* 1995, 108:961-968.
84. Davies Jr., F.T., Calderon, C.M. et al. *Scientia Horticulturae* 2005, 106:318-329.
85. Borisov, A.Y., Danilova, T.N., Koroleva, T.A. et al. *Fundamentals and Application* 2004, 13:137-144.
86. Jacobi, L.M., Zubkova, L.A., Barmicheva, E.M. et al. *Mycorrhiza* 2003, 13:9-16.
87. Liu, H., Trieu, A.T., Blaylock, L.A. et al. *Molecular Plant-Microbe Interactions* 1998, 11:14-22.
88. Kim, K.Y., Jordan, D. and Mc Donald, G.A. *Biology and Fertility of Soils* 1998, 26:79-87.
89. Catoira, R., Galera, C., De Billy, F. et al. *Plant Cell* 2000, 12:1647-1666.
90. Cook, D.R. *Current Opinion in Plant Biology* 1999, 2:301-304.

2

Ectomycorrhizae and Forest Plants Fitness

L. Montecchio and L. Scattolin

Dipartimento Territorio e Sistemi Agro-Forestali
Università degli Studi di Padova, viale dell'Università 16
35020 Legnaro, Padua, Italy

Introduction

According to the most recent information, mycorrhizal symbiosis is expressed within a wide gradient that, while having extremes of parasitism of fungus on plant and parasitism of plant on fungus, commonly consists of mutualism between the two bionts [1].

The ectomycorrhizae (EMs), found mainly on the forest tree species of temperate climes and, to a lesser extent, on shrubs and herbaceous plants, involve the majority of the living root tips of a plant, with high fungal species richness and a mutable presence and frequency over time. This is the result of a complex and dynamic sequence of relationships [2-5], mainly influenced by the characteristics of the two partners (plant and fungus), the interactions between EM and other symbionts of the same plant, the biotic and abiotic mycorrhizospheric features [6-10], site features [11-16], and health status of the plant [17-19].

Within this perspective, the dynamics controlling the persistence and competition between EM fungi and, consequently, the EM community structure at both plant and population level, play an important role [20] in contributing to the maintenance of plant fitness within the wider context of forest ecosystem, also when natural and human disturbances occur [21-23].

EM Persistence and Competition among EM Fungi

EM are ephemeral structures. This means that, in order to persist over the years, a fungus must be able to move from the tip originally colonised. This may follow two mechanisms: (a) diffusion in parallel with the growth of the fine root to which the fungus is associated, and successive colonisation of its newly-formed rootlets; and (b) colonisation of new root tips in yet unexplored areas of the root system.

EM tips normally grow slowly, and their emerging ramifications are rapidly colonised by the fungal component mainly belonging to the mantle [5]. Other root tips in different areas of the same root system may be colonised by the growth of the extramatrical mycelium. In this way the diffusion of the fungus will depend on its growth rate in relation to that of the root tips and on the presence and distribution of antagonistic microorganisms in the soil.

The success of one mycorrhizal fungus over another can be achieved by two possible mechanisms: competitive exclusion and antagonistic exclusion. There is competitive exclusion when a fungus demonstrates a greater capacity to colonise thanks to its faster growth rate, higher density, position of its inoculum or, lastly, a greater affinity for the host, soil or other environmental conditions. The existence of relationships of antagonistic exclusion has not yet been demonstrated. They would imply the capacity of a fungus to colonise the host thanks to its antagonistic behaviour towards other fungi competitors.

Widening the concept to the fungal community, variants can be added such as competitive and antagonistic substitution. There is competitive substitution, in space, when one fungus more efficient than another manages to colonise a new part of the root system and use the nutrients obtained by the host to develop mycelium, that acts as a further inoculum [25-27] or, in time, when substantial changes in the substrate (e.g. progressive aging of the plant partner) favour the better-adapted species in that mutable trophic context [28]. Antagonistic substitution takes place when one fungus is able to substitute another by direct antagonistic action against it (antibiosis [29]).

EM Community at Plant Level

It is difficult to identify a joint role of the EM within a community: a single plant may host an enormous number of fungal symbionts, and the physiological differences among species influence the environment in various ways [30].

Many points still await clarification: their structure in terms of species number and abundance; the ways in which factors such as specificity, differences in the soil, organic matter supply, successions and disturbances act on them; the magnitude and mechanisms with which the EM communities act on the functions of the ecosystem and structure of the plant communities and the autoecology of the more important fungal species [31].

Generally, in a root system a few fungal taxa represent the majority of EM presence, with many other species being relatively infrequent [16, 17, 32, 33]. Moreover, the EM species present even in small volumes of soil have a highly heterogeneous distribution both horizontally [31, 34-36] and vertically [10, 37, 38]. The spatial distribution also varies over time [39], as different EM mycelia show different persistence and ways of growing and reproducing. This is manifested in the formation of few very large *genets,* that can densely occupy wide aggregated areas for a long time [40], or in numerous but small mycelia that are not very persistent and scattered in the community [41, 42].

The EM effect on the plant also depends on the particular characteristics of a fungus with respect to another, as each fungus shows particular aptitudes in given environmental conditions or when in symbiosis with given plant species [43-46], expressing a higher capacity to absorb a specific element or to resist drought conditions, higher resistance to the presence of toxic heavy metals, or high specificity for the host, etc. Furthermore, there can be marked variability among strains of the same species, as a result of a different adaptation to various soils and host plants [47].

All these elements can strongly influence the competitive interactions between plants. For example, *Rhizopogon vinicolor* confers greater drought resistance on seedlings of *Douglasia* than other fungi [48], *Laccaria laccata* survives much longer on the roots of cut seedlings than *Hebeloma crustuliniforme* [49], and *H. crustuliniforme* exhibits higher efficiency in extracting N from proteins than *Amanita muscaria* or *Paxillus involutus* [50].

The EM fungal succession, the EM plant age and physiological state, and the degree of specificity of the fungus for the plant can influence the competition among plants. The less selective fungi benefit from the greater probability of finding hosts with which to associate, whereas more specific fungi, which coevolved for a long time with the same partner, may exhibit higher compatibility and confer greater competitivity on the plant than the less specialised wide-spectrum fungi [30, 51].

Recently, a study performed in a birch (*Betula pendula*) woodland in spontaneous expansion on an abandoned alpine meadow demonstrated the presence of a specific mycorrhizal consortium in the plants at the forefront of woodland expansion, supporting the hypothesis that the colonisation of non-forestry lands by a woodland tree is mediated by the presence and interaction of pioneer symbiont species [52].

The deep bond between root tips and EM fungi, the high complexity of this hidden ecosystem and the difficulty in exploring and describing it, therefore, make it necessary to improve and standardize methods to study the aboveground dynamics and their relations with forest tree physiology. In this context, optimized samplings and experimental designs would give a good description of the plant root system not only from a physiological point of view, but also from an ecological and phytosanitary one. According to Lilleskov and Parrent [53], it would also provide a baseline against which to measure the effect of future environmental change, allowing the researcher to determine where and how fungal communities are responding to variable resources and conditions.

Classical statistical methods for representing environmental variables often assume that measurements are uniformly or randomly distributed, but this concept is often inappropriate for analyses of several environmental measures, such as the different kind of organisms (i.e. bacterial or fungal communities, pedofauna) or soil, water and air contents: values at neighbouring locations are in fact rarely independent, particularly over short distances. This kind of dependence, called spatial autocorrelation, nonetheless makes it possible to interpolate values at unmonitored locations from known values at monitored locations. New tools and technology are helping to improve our knowledge. One interesting branch of applied statistical methodology, based on the modelling of spatial relationships and dependence, is geostatistics: used for modelling spatial data, it provides accurate estimations of phenomena at locations for which no measurements are available. The main concept in using geostatistics to characterize spatial heterogeneity is that environmental features are not random, but have some spatial continuity correlated over a given distance [54]. The capacity to describe systems in a more appropriate way that is closer to reality has led to an increasing use of geostatistics and geographic information systems (GIS) in forestry and agricultural research [55].

Geostatistics supply a set of tools useful for characterizing variability in space. By means of the kriging method (synonymous with 'optimal prediction', according to Journel and Huijbregts [56]), it is possible to take the spatial autocorrelation structure function (variogram) into account by means of known values from monitored locations, weighting them with values read from the variogram at corresponding distances, and splitting weights among adjacent locations [57, 58]. A basic aspect of the kriging method, which weighs the surrounding measured values to derive a prediction for an unmeasured location, is that the weights are based not only on the distance between the measured points and the location to be predicted, but also on the overall spatial arrangement of the measured points [59]. Recent researches suggested that spatial gradients of the examined features exist at plant level, associated to the up-downslope direction and distance from the stem. The effectiveness of the geostatistical model used demonstrated that a geometrical sampling design associated to spatial mapping techniques can be useful in research where the tree, and not the forest, is the subject [16, 60, 61].

EM at Stand Level

EM communities are highly complex, with many different factors that influence their dynamics in various ways. It is thus not possible to talk about a joint 'mycorrhizal effect' [2]. The EM can influence the plant community with their extramatrical mycelium, connecting the root systems of different plants and mediating the exchange of nutrients [30, 48, 62-65]. Although it had been known for some time that mycorrhizal fungi linked the root systems of different plants, until the experiment by Simard et al. [66] it was unclear if there was a quantitatively significant transfer of nutrients over these 'mycelial bridges' [5]. It has now been ascertained that, at least as regards organic carbon, this phenomenon exists [66, 67].

Perry et al. [68] stated that the presence of these links in EM communities led to the formation of true 'plant corporations' or 'associations for mutual aid and the promotion of common interests'. Numerous subsequent observations have confirmed the presence of these connections and their transfer of carbon between plants [62, 68-71]. However, this condition flouts the more modern 'individualist' theory of competition formulated by Tilman [72, 73], according to which organisms are in 'conflict' over the use of available resources, and that the plant more capable of exploiting the resources would grow to the detriment of those less able. According to this theory, the competitively weaker plant would be unable to oppose the sink caused by the stronger plant, so would be unable to divert resources towards itself. The transfer of nutrients could thus only be conceived of when the resource was present in non-limiting amounts. In this case the surplus produced by a plant might be transferred to another weaker one [48]. Therefore, these arguments only remain valid if the EM is considered as a simple passive link between the two plants. On the contrary, if it is seen as something that can actively control the flow of nutrients through its mycelium network, the facilitation or competition would no longer be mechanisms occurring directly between plants, but would come from the relationship either between the plant and its fungal symbiont, or between the two plants but with the active participation of the shared fungus [48].

Nutrient Recycling in the System through Mycelium Networks

Another way in which the presence of the mycelial network might influence the competition among plants is the possibility that the fungal symbionts can recycle nutrients in the system which would otherwise be lost following the death of one or more plants. In this case the fungus is in a privileged position for the uptake of nutrients released by the dead tissues, but it can only recycle these nutrients if one of the two situations described below occurs.

The first is that the EM and extramatrical mycelium manage to survive for a certain period after the death of the plant, even without a nutrient supply from nearby living plants, and can then colonise new seedlings within the time window during which the fungal structures can survive without an energy supply (for example, following the clear cutting of a woodland and its subsequent renewal). It has been demonstrated that EM are able to survive the death of the root or plant at least for a given period and that during this time there is only a slight loss of nutrients from the system [74, 75]. The second circumstance is that the EM of the dead plant remains linked by its mycelium network to other living plants nearby; it receives nourishment from these latter and, at the same time, immediately recycles the substances released by the tissues of the dead plant within the system and thus avoids losses of N through leaching. However, the recycling of nutrients by this mechanism is only possible below a given threshold of adjacent cut trees. Above that threshold, the lack of continuity prevents the mycelium network maintaining the links [48, 76].

These mechanisms can affect the competition between plants as they determine a higher nutrient availability in the system, which may produce less competition [48]. In conclusion, the phenomena linked to the flow of nutrients through the mycelium network can influence the competition between plants by means of five possible mechanisms [77]:

- new seedlings can rapidly be connected to the mycelium network
- dominant plants can 'donate' carbon to those relegated to a subordinate position
- the link can balance the competition between plants
- mineral nutrients can pass from one plant to another
- nutrients from dying plants can be transferred to living plants through the fungal mycelium.

EM and the Functioning of Forest Ecosystems

Studies on EM communities have often been mainly correlative, rarely covering or describing the mechanisms underlying the dynamics of these communities [78]. Numerous fungi probably perform similar functions, with a certain degree of functional redundancy existing in EM communities. Scattolin et al. [16], for example, in a study on EM communities in *Picea abies*, strengthened the hypothesis that marked changes in the ectomycorrhizal community depend on the selective pressure of some environmental features [79], suggesting that bedrock pH and exposure, interfering with wider environmental factors involving root system development and plant nutrition, play a crucial role on both tips' turn-over and EM status, in accordance with well-known colonisation strategies [78, 80-82]. Soil pH and site exposure could therefore play a primary role in the adaptive selection of species or functional groups, both directly, acting on the tolerance of the fungal species (or genotypes), and indirectly, through dynamics involving plant nutrition, where nutrient availability and translocation could be essential [2, 38, 82, 83]. Indeed, soil fungi are known to be biogeochemical agents that can influence weathering through physical and chemical processes [84-88].

Direct weathering and nutrient uptake by ECM fungi colonising mineral particles have also been suggested as a possible pathway for element uptake by forest trees [86, 89]. Although the quantitative importance of fungal weathering in plant nutrition remains controversial [90], the rate and composition of the ECM community could change in terms of contributing to tree nutrition. According to Courty et al. [91], this aspect could engage a variety of mechanisms for mobilizing nutrients from the soil, involving enzymatic degradation of macromolecules, metal complexation and mineral weathering.

From a functional point of view, therefore, it can be supposed that the fungal communities adapted to different sites can, on the whole, mobilize nutrients essential to trees in a comparable way. For instance, assuming that similar amounts of nitrogen are available to the trees in the investigated stands, it can be expected that when organic nitrogen sources in the soil profile are unsatisfactory, two different ECM consortia distinctive of acid and basic plots could play an active role in making N available from ammonium and nitric ions, respectively, using different strategies [16].

Hunt and Wall [92] observed that the functional groups of soil ecosystems that have most impact on the others are those with the highest biomass. They also established that ecosystems can support the loss of some functional groups without this prejudicing their general functioning. However, this is probably due to the intervention of as yet unknown stabilising mechanisms in the system. The authors hope to produce theoretical models that are able to predict the effects of a loss of biodiversity on ecosystems.

A possible species redundancy should therefore not lead to underestimating the importance of biodiversity in ecosystems, as this increases the probability that species exist which can perform important functions if environmental or climatic changes of a certain magnitude occur. Diversity could thus be a form of 'environmental capital natural insurance' [93]. If it is difficult to predict which species would be important in altered environmental conditions [92, 94], it would be even more hazardous to try to predict future environmental changes [94]. Methods exist to evaluate the action of species which potentially perform the function of 'key' species, for example the most abundant EM fungi. Testing the groupings of these fungi in controlled environmental situations, by applying different parameters it should be possible to evaluate their function and understand the effects that these communities have on the nutrient cycle, carbon balance and the response to disturbances [17, 32].

Among the many variables that may contribute to the segregation of EM species [30, 38, 82, 83], two particularly important occurrences are drastic silvicultural practices (i.e. clear cuttings [95-97]) and unpredictable natural disturbances, such as disease driven by abiotic factors (i.e. forest decline [17]).

Sylviculture and EM Communities

Sylvicultural practices have a strong influence on the quantity and quality of the EM fungi population as they usually involve very intense disturbance to the system [96, 98-102]. Given the importance of these fungi in the dynamics of forest communities, it is clear that an understanding of the effects that different cultural practices may have on the EM fungi is of major interest in order to be able to apply the correct techniques. Considering the intense levels of exploitation in the majority of the world's forests, the disturbances caused by sylvicultural practices can be equated to those of a natural type and, like them, may determine selection pressures over time that favour a given type of species composition in the EM communities [96].

Forestry practices can directly influence the fungal populations or modify the effects of other types of disturbances. Byr et al. [99], for example, compared the composition of EM communities in century-old forests of *Pinus contorta* and sites that had been clearcut in the forests of Montana (USA). The effect of clearcutting was a reduction in the overall number of species of EM fungi, while the relative species abundance were similar to those of the undisturbed sites.

An important aspect is the possibility that there might be a long-term reduction in the number of species of EM fungi because this would affect the resilience of the ecosystem to subsequent disturbances [32, 94, 99]. Indeed, although many believe that functional redundancy probably exists in EM fungi communities and it is therefore likely that the system could remain stable and functional with fewer fungal species, it is also true that current knowledge on the functions of the different species is too limited for an evaluation of the effect that the loss of one or more species might have on the ecosystem in the short-term and even more so in the long-term. Added to this, is the unpredictability of potential future disturbances to the system, so it must be concluded that the conservation of biodiversity is the only logical approach to adopt [32, 94].

Clearcutting, through its negative effect on the EM fungi, also leads to a loss of nutrients from the system. This was demonstrated by Parsons [76], and is described in the chapter on plant competition. Mahmood et al. [102] observed that the systematic removal of logging residues for use as an alternative to fossil fuels leads to a reduction in soil organic matter and the colonisation rate of root tips by EM fungi. Amaranthus and Perry [98] ascertained that woody residues in the forest soil are also very important for the conservation of EM fungi following a fire. After the fires in southern Oregon in 1987, most of the inoculum available for the subsequent re-colonisation was concentrated in the decomposing wood. During dry periods, decomposing wood can preserve moisture and thus be a suitable environment for the survival of the inoculum. Leaving woody residues on site or even distributing it might therefore be a way to increase the resilience of a woodland [103].

Sylvicultural practices that cause soil compaction, such as those using heavy machinery, also have an inhibitory effect on the growth of fungi [103]. As previously mentioned, EM can survive for a time after a plant has been cut, but this period is limited and if host plants are not introduced to the site within this period, the fungal inoculum will be lost. Any invasion of the site by non-EM plants, because of their negative effect on inoculum survival, would jeopardise the possibilities of regeneration of the woodland, especially in those difficult environmental situations where plants require rapid colonisation to be able to survive [103]. The conservation of EM species of the undergrowth is a useful way to keep a fungal inoculum alive after the trees have been removed, allowing new tree seedlings to take root and grow much faster.

Forest Decline and EM Communities

Oak decline has been a well-documented phenomenon in Europe from the beginning of the 20th century, and since the 1980s has become a problem worldwide, spreading through Europe, Asia, North and South America [104, 105]. It was first reported in Germany and Switzerland in 1739 and 1850, respectively, then in France in 1875 and 1893 [106]. Subsequently in the 1920s, Europe was stricken by a new wave of oak decline; in Italy the regions that most suffered were the Po Plain and Apennine Mountains, where many trees aged around 50-60 years declined and died. Decline has a complex aetiology due to the dynamic interactions between the host plant and several abiotic and biotic causes. Manion [107] proposed a model with three categories of factors according to their action: the *predisposing* factors, acting in the long-term and weakening tree vitality (i.e. tree genetic characteristics, environmental changes), the short term *inciting* factors (i.e. frost, drought, excessive plant density and competition), with a more intense action, leading up to the appearance of decline symptoms, and the *contributing* factors (i.e. fungal parasites and bark beetles), acting in the final stages of decline and emphasising it [108, 109].

Decline symptoms are visible both at canopy and trunk levels. Canopy transparency gradually increases because of the yellowing, wilting and fall of leaves, bud abscission, and reduced and delayed new leaves growth. The internodes then shorten, and numerous epicormic twigs appear along the trunk and close to the collar [17, 110]. In severely declining trees the branches also dry out, longitudinal cracks in the bark appear along the trunk, and several fungal parasites can be detected in both the trunk and root system (*Biscogniauxia* sp., *Diplodia* sp., *Collybia* sp., *Armillaria* sp., *Phytophthora* sp.) [111-113]. Death may also be hastened by weakness caused by fungi that are usually endophytic and asymptomatic on healthy plants [114, 115].

Studies on many forest species have demonstrated that the fine roots of declining trees often show functional and anatomical anomalies and changes of the ectomycorrhizal status [116-120]. Holm oak (*Quercus ilex* L.) decline symptoms are more intense where trees are a long way from the surface water and the groundwater, and where salts concentration is higher, suggesting that drought and water salinity could play an important role as environmental factors able to predispose the less stress-resistant genotypes to decline [81, 109, 120-122].

Results obtained by Montecchio et al. [17] demonstrated that the relative frequency of some EM is related to the decline intensity, and that the most frequent EMs (27%) relative frequencies strongly discriminated among asymptomatic, weakly and strongly declining trees (Figure 1), indicating the possibility of characterising the decline degree also through below-ground features, studying the recovery frequency of the more frequent ectomycorrhizae

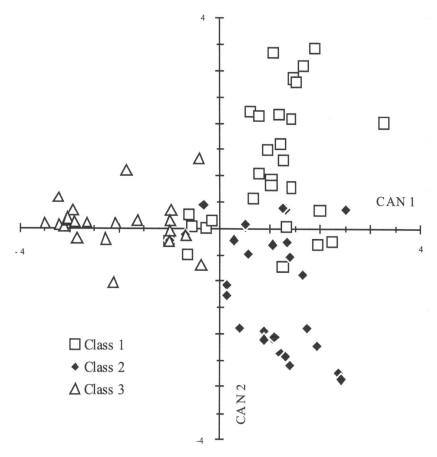

Figure 1: Canonical discriminant analysis of the more frequent morphotypes (P<0.0001) in a declining holm oak forest. Class 1= healthy and asymptomatic plants, class 2= weakly declining plants; class 3 = strongly declining plants (five plants/class).

independently from the rarest ones. Furthermore, only one third of the detected EM were present in all decline classes, while the occurrence and abundance of those assigned to a particular class changed with the decline status, increasing in number and decreasing in frequency. The results of this study, obtained from trees at different decline levels growing in close proximity to each other in the same environment, therefore, strengthen the hypothesis that adjustment in the EM community is an indirect effect of the selective pressure of the environmental stresses on the tree and its rootlets. Depending on an individual susceptibility, the tree gradually declines maintaining its full ectomycorrhization, but losing progressively the ability to select the most efficient fungal symbionts and allowing their substitution with others.

According to well-known colonisation strategies, both the composition of the fungal population and its relative abundance could gradually change with the medium [78-81, 123]. In this hypothesis, while the EMs common to both asymptomatic and declining trees could have a wider adaptive range, the few ectomycorrhizae peculiar to asymptomatic trees could be gradually replaced by an increasing amount of others, essential to maintain the stability of the modified ecosystem processes but probably less efficient than the previous ones [124, 125]. Examples of ectomycorrhizal fungi that do not act in a mutualistic manner throughout their entire life cycle (colonizing the tip as weak saprophytes or parasites, or able to grow saprophytically independently from the tip, or forming resting stages) are well known [126-131] and the saprophytic or resting ability of some fungi characterising the most declining trees, rapidly growing as mycelium or producing sclerotia on artificial media are known from laboratory observations.

Similar results were obtained in a thinned declining Common oak (*Quercus robur*) forest [97], where the sylvicultural treatment altered the relative EM species abundance in the community mainly according to the health status of the plants. Prior to thinning, the ectomycorrhizal community structure differed in the two decline classes, with higher biodiversity in the less declining trees. Eighteen months later the difference between decline classes persisted only in the untreated trees, while it was no longer significant in the treated ones. At the same time, the distinction between untreated and treated trees belonging to the severe decline class was significant, demonstrating that the relative abundance of one third of the whole EMs can discriminate between treated and untreated trees. Furthermore, one third of the EMs detected were present in both decline classes, while the occurrence and abundance of those assigned to a particular class changed with the decline status, increasing in number and decreasing in frequency.

Conclusion

In general, it can be stated that EM contribute towards modifying competitivity in plant communities, tending to facilitate those plants that respond better to colonisation. These are often the more mature plants. The EM presence entails an overall increase in plant biomass and in species diversity and uniformity [132, 133].

The flow of nutrients between plants through the fungal mycelia can regulate and reduce the competition between them. This means, for example, that a seedling growing in the shade of the tree canopy can receive photosynthates from the adult plants [66, 67]. The possibility of a rapid recycling of nutrients and avoiding their loss from the system could be another way in which the fungal community reduces competition between plants by means of an overall increase in the availability of these elements. Competition is also reduced by the capacity of the fungi to segregate into a large number of niches through a complex series of mechanisms. Although much has still to be learned about the mechanisms with which the mycorrhizal fungi regulate competition between plants, it is certain that they play a fundamental role in the fitness of a plant and the forest consortium.

Intensive exploitation of the environmental resources is required by an increasing world population, together with the economic growth of developing nations, and will probably be required even more so in the future to meet the resulting demands. Given the scarcity of available 'natural capital', use of the ecosystems for production needs must be made in a way that their productivity and health are not compromised, but maintained in the long-term.

Stable natural environments are often characterised by the presence of organisms with high resistance to disturbance, but an equally low resilience renders re-colonisation difficult after a disturbance that was severe enough to eliminate these plants from the habitat [135]. This means that if an unexpected disturbance (disease, logging) passes certain levels in terms of intensity and area, the system risks rapidly losing its capacity for recovery, at least in the short-term [2, 135]. There is, therefore, a need to formulate techniques and criteria for the exploitation of resources that derive from a thorough knowledge of all the mechanisms regulating these complex natural systems in order to avoid exceeding the critical thresholds for their survival.

It is also probable that there will shortly be a more frequent need to apply operations of environmental restoration, including on a vast scale, where a 'mindless' use of the natural capital has produced its irreversible degradation. Almost all ecosystems are suffering from a rapid loss of their overall biodiversity [136]. It is not yet clear how this loss will influence their productivity and stability and to what extent the functional redundancy existing in these systems can compensate for the reduction in species.

It has been hypothesised that systems could remain functioning after the loss of many species or even of an entire functional group [136, 137]. However, this should not lead to an underestimation of the implicit risks in this process. According to Bengtsson [94], if it is theoretically possible to be able to predict in detail the effects that the loss of one or more species might have on a system in given conditions, it would be arrogant to think of predicting all the possible environmental conditions that might be created in the future. The conservation of biodiversity will allow the 'capital natural insurance' that it represents to be maintained intact.

The conviction is growing that the communities of soil organisms have enormous importance in the functioning of terrestrial ecosystems [138], and in most of these communities the EM form the most significant portion of biomass and play a primary role in influencing the nutrient cycle [9, 31, 138, 139].

Study of the EM should therefore be considered a necessary step to reach a full understanding of the mechanisms involved in the functioning of the different terrestrial systems. The presence of this symbiosis can in no way be ignored within the framework of compatible resource management. Although studies on this phenomenon have multiplied, especially over the last twenty years, they have so far contributed more to creating an awareness of the importance of this symbiosis rather than an understanding of the complex mechanisms that regulate its functioning and interactions [9, 31, 138].

Practical applications already exist of the knowledge that has accrued on EM: they have been used to accelerate succession and obtain faster and more reliable restoration of particularly disturbed environments [140, 141]. The use of artificially inoculated plants for the introduction of exotic trees has become standard practice. The colonising of native plants in highly disturbed soils also benefits enormously from prior inoculation [5]. Suitable fungal strains have been selected and marketed for this purpose, in order to guarantee plant rooting and faster growth. However, these practices have drawbacks in the cases where native species are introduced in not excessively disturbed sites or close to still intact areas. Where a sufficiently viable natural fungal population already exists, complex interactions take place that modify and often annul the effect of these selected strains on the plants [142].

Even without understanding exactly how these mechanisms work, it is obvious that the end result of their action has a basic effect on the system equilibrium. Consequently, it is equally obvious that improved knowledge on these phenomena is a necessary condition for understanding an ecosystem's potential and capacity for recovery and therefore, in a final analysis, for an appropriate use of the territory. There is a tendency to correlate species diversity *per se* to ecosystem functionality, but this might not be entirely accurate, because the combination of members of these communities exhibits high functional redundancy and many of the functions are performed by a few key species, possibly the most abundant ones, even when considering the probable role of those less frequent as a limiting factor. The identification and understanding of the characteristics and role of these species must be improved in future studies.

References

1. Schulz, B. and Boyle, C. *Mycol. Res.* 2005, 109:661-686.
2. Allen, M.F. The Ecology of mycorrhizae. Cambridge University Press, Cambridge, 1991, p. 184.
3. van der Heijden, M.G.A. and Sanders, I.R. Mycorrhizal Ecology. Springer-Verlag, Berlin, Germany, 2002, p. 469.
4. Simard, S.W. and Durall, D.M. *Can. J. Botany* 2004, 82:1140-1165.
5. Smith, S.E. and Read, D.J. Mycorrhizal Symbiosis, 2nd ed., Academic Press, London, 1997, p. 605.

6. Dickie, I.A., Xu, B. and Koide, R.T. *New Phytol.* 2002, 156:527-535.
7. Koide, R.T., Xu, B., Sharda, J. et al. *New Phytol.* 2005, 165:305-316.
8. Lilleskov, E.A., Fathey, T.J. and Lovett, G.M. *Ecol. Appl.* 2001, 11:397-410.
9. Read, D.J. and Perez-Moreno, J. *New Phytol.* 2003, 157:475-492.
10. Rosling, A., Landeweert, R., Lindahl, B.D. et al. *New Phytol.* 2003, 159:775-783.
11. Conn, C. and Dighton, J. *Soil Biol. Biochem.* 2000, 32:489-496.
12. Courty, P.E., Pouysegur, R., Buee, M. et al. *Soil Biol. Biochem.* 2006, 38:1219-1222.
13. Dickie, I.A. and Reich, P.B. *J. Ecol.* 2005, 93:244-255.
14. Gehring, C.A., Theimer, T.C., Whitham, T.G. et al. *Ecology* 1998, 79:1562-1572.
15. O'Dell, T.E., Ammirati, J.F. and Schreiner, E.G. *Can. J. Bot.* 1999, 77:1699-1711.
16. Scattolin, L., Montecchio, L. and Agerer, R. *Trees - Struct. Funct.* 2008, 22:13-22.
17. Montecchio, L., Causin, R., Rossi, S. et al. *Phytopathol. Mediterr.* 2004, 43:26-34.
18. Mosca, E., Montecchio, L., Sella, L. et al. *Forest Ecol. Manag.* 2007, 244:129-140.
19. Pfleger, F.L. and Linderman, R.G. Mycorrhizae and Plant Health. APS Press, St. Paul-MN, 1994, p. 344.
20. Goodman, D.M. and Trofymow, J.A. *Can. J. Forest Res.* 1998, 28:574-581.
21. Lepšovà, A. and Mejstøik, V. *Nature* 1992, 28:305-312.
22. Reid, C.P.P. *In:* Mycorrhizas in the Rhizosphere. Lynch, J.M. (ed.), Wiley-Interscience, Chichester, England, 1990, pp. 281-315.
23. Tennakoon, I.A., Gunatilleke, U.N., Hafeel, K.M. et al. *Forest Ecol. Manag.* 2005, 208:399-405.
24. Bowen, D.J. and Theodoru, C. *In:* Ectomycorrhizae. Marks, G.C. and Kozlowski, T.T. (eds.), Academic Press, New York, 1973, pp. 107-150.
25. Cooke, R.C. and Rayner, A.D.M. Ecology of saprophytic fungi, Cambridge University Press, Cambridge, 1984.
26. Fleming, L.V. *Plant Soil* 1983, 71:263-267.
27. Fleming, L.V. *New Phytol.* 1984, 98:143-153.
28. Giltrap, N.J. *T. Brit. Mycol. Soc.* 1982, 78:75-81.
29. Marx, D.H. *Phytopathology* 1970, 60:1472-1473.
30. Molina, R., Massicotte, H. and Trappe, J.M. *In:* Mycorrhizal Functioning. Allen, M.F. (ed.), Chapman and Hall, London, 1992, pp. 357-423.
31. Horton, T.R. and Bruns, T.D. *Mol Ecol.* 2001, 10:1855-1871.
32. Dahlberg, A. *New Phytol.* 2001, 150:555-562.
33. Jonsson, T., Kokalj, S., Finlay, R.D. et al. *Mycol. Res.* 1999, 103:501-508.
34. Baier, R., Ingenhaag, J., Blaschke, H. et al. *Mycorrhiza* 2006,16:197-206.
35. Dahlberg, A., Jonsson, L. and Nylund, J.E. *Can. J. Botany* 1997, 8:1323-1335.
36. Genney, D.R., Anderson, I.C., Alexander, I.J. *New Phytol.* 2006, 170:381-390.
37. Scattolin, L., Montecchio, L., Mosca, E. et al. *Eur. J. For. Res.* 2009 (in press).
38. Tedersoo, L., Koljalg, U., Hallenberg, N. et al. *New Phytol.* 2003, 159:153-165.
39. Koide, R.T., Shumway, D.L., Bing, X. et al. *New Phytol.* 2007, 174:420-429.
40. Bonello, P., Bruns, T.D. and Gardes, M. *New Phytol.* 1998, 133:533-542.
41. Redecker, D., Szaro, T.M., Bowman, R.J. et al. *Mol. Ecol.* 2001, 10:1025-1034.

42. Guidot, A., Debaud, J.C., Effosse, A. et al. *New Phytol.* 2004, 161:539-547.
43. Gorissen, A. and Kuyper, Th.W. *New Phytol.* 2000, 146:163-168.
44. Helgason, T., Merryweather, J.W., Denison, J. et al. *J. Ecol.* 2002, 90:371-384.
45. Koide, R.T. *In:* Arbuscular mycorrhizas: Physiology and function. Kapulnick, Y. and Douds, D.D. (eds.), Kluwer Academic Press, 2000, pp. 19-46.
46. Marulanda, A., Azcóna, R. and Ruiz-Lozanoa, J.M. *Physiol. Plantarum* 2003, 119:526-533.
47. Tibbett, M., Sanders, F.E. and Sanders, M. *Mycol. Res.* 1998, 103:707-714.
48. Miller, S.L. and Allen, E.B. *In:* Mycorrhizal Functioning. Allen, M.F. (ed.), Chapman & Hall, New York, 1992, pp. 301-332.
49. Parke, J.L., Linderman, R.G. and Black, C.H. *New Phytol.* 1983, 95:83-95.
50. Abuzinadah, R.A. and Read, D.J. *New Phytol.* 1989, 112:61-68.
51. Brundrett, M.C. *New Phytol.* 2002, 154:275-304.
52. Scattolin, L., Montecchio, L. and Mutto Accordi, S. *Micologia Italiana* 2007,1:58-66.
53. Lilleskov, E.A. and Parrent, J.L. *New Phytol.* 2007, 174:250-256.
54. Sarangi, A., Madramootoo, C.A., Enright, P. et al. *Water, Air, Soil Pollut.* 2005, 168:267-288.
55. Nelson, M.R., Orum, T.M., Jaime-Garcia, R. et al. *Plant Dis.* 1999, 83:308-319.
56. Journel, A.G. and Huijbregts, CH. J. 1981. Mining Geostatistics. Academic Press, London, p. 600.
57. Liao, D., Peuquet, D.J., Duan, Y. et al. *Environ. Health Perspect.* 2006, 114:1374-1380.
58. Webster, R. and Oliver, M.A. Geostatistics for Environmental Scientist. John Wiley & Sons, Chichester-UK, 2001, p. 271.
59. Johnston, K., Ver Hoef, J.M., Krivoruchko, K. et al. Using ArcGIS Geostatistical Analyst. ESRI Inc., Redlands-CA, 2001, p. 300.
60. Mosca, E., Montecchio, L., Scattolin, L. et al. *Soil Biol. Biochem.* 2007, 39:2897-2904.
61. Scattolin, L. and Montecchio, L. *Pl. Dis.* 2007, 91:771.
62. Fitter, A.H., Graves, J.D., Watkins, N.K. et al. *Funct. Ecol.* 1998, 12:406-412.
63. Grime, J.P., Mackey, J.M., Hillier, S.H. et al. *Nature* 1987, 328:420- 422.
64. Kennedy, P.G., Izzo, A.D. and Bruns, T.D. *J. Ecol.* 2003, 91:1071-1080.
65. Sims, S., Hendricks, J., Mitchell, R. et al. *Mycorrhiza* 2007, 17:299-309.
66. Simard, S.W., Perry, D.A., Jones, M.D. et al. *Nature* 1997, 338:579-582.
67. Simard, S.W., Molina, R., Smith, J.E. et al. *Can. J. Forest Res.* 1997, 27:331-342.
68. Perry, D.A., Margolis, H., Choquette, C. et al. *New Phytol.* 1989,112:501-511.
69. Jones, M.D., Durall, D.M., Harniman, S.M.K. et al. *Can. J. Forest Res.* 1997, 27:1872-1889.
70. Massicotte, H.B., Molina, R., Tackaberry, L. et al. *Can. J. Botany* 1999, 77:1053-1776.
71. Finlay, R.D. and Read, D.J. *New Phytol.* 1986, 103:143-156.
72. Tilman, D. Resource Competition and Community Structure. Princeton University Press, Princeton, 1982, p. 296.
73. Tilman, D. Plant Strategies and the Dynamics and Structure of Plant Communities. Princeton University Press, Princeton, 1988, p. 360.

74. Al Abras, K., Bilger, I., Martin, F. et al. *New Phytol.* 1988, 110:535-540.
75. Marshall, J.D. and Perry, D.A. *Can. J. Forest Res.* 1987, 17:872-877.
76. Parsons, W.F.J., Miller, S.L. and Knight, D.H. 1990. *Bull. Ecol. Soc. Am.* 1990, 71:280.
77. Newman, EI. *Adv. Ecol. Res.* 1988, 18:243-270.
78. Lilleskov, E.A. and Bruns, T.D. *New Phytol.* 2001, 149:154-158.
79. Agerer, R. *Mycorrhiza* 2001, 11:107-114.
80. Grayston, S.J. and Campbell, C.D. *Tree Physiol.* 1996, 16:1031-1038.
81. Shi, L.B., Guttenberger, M., Kottke, I. et al. *Mycorrhiza* 2002, 12:303-311.
82. Taylor, A.F.S. and Alexander, I.J. *Agricult. Ecosys. Environ.* 1989, 28:493-497.
83. Deacon, J.W. and Fleming, L.V. *In:* Mycorrhizal functioning: An integrative plant-fungal process. Allen, M.F. (ed.), Chapman and Hall, New York, 1992, pp. 249-300.
84. Banfield, J.F., Barker, W.W., Welch, S.A. et al. *Proc Natl Acad Sci USA* 1999, 96:3404-3411.
85. Bornysaz, M.Z., Graham, R.C. and Allen, M.F. *Geoderma* 2005, 126:40-160.
86. Landeweert, R., Hoffland, E., Finlay, R. et al. *Trends Ecol. Evolut.* 2001, 16:248-254.
87. Schenk, H.J. *New Phytol.* 2008, 178:223-225.
88. Sterflinger, K. *Geomicrobiol.* 2000, 17:97-124.
89. Rumberger, M.D., Münzenberger, B. and Bens, O. et al. *Plant Soil* 2004, 264:111-126.
90. Sverdrup, H., Hagen-Thorn, A., Holmquist, J. et al. *In:* Developing principles and models for sustainable forestry in Sweden. Sverdrup, H. and Stjernquist, I. (eds.), Kluwer Academic Publishers, Dordrecht, 2002, pp. 91-196.
91. Courty, P.E., Pritsch, K., Schloter, M. et al. *New Phytol.* 2005, 167:309-319.
92. Hunt, H.W. and Wall, D.H. *Global Change Biol.* 2002, 8:33-50.
93. Folke, C., Holling, C.S. and Perrings, C. *Ecol. Appl.* 1996, 6:1018-1024.
94. Bengstsson, K. *Appl. Soil Ecol.* 1998, 10:191-199.
95. Bruns, T.D., Bidartondo, M.I. and Taylor, D.L. *Integr. Comp. Biol.* 2002, 42:352-359.
96. Jones, M.D., Daniel, M.D. and Cairney, J.W.G. *New Phytol.* 2003, 157:399-422.
97. Mosca, E., Montecchio, L., Sella, L. et al. *Forest Ecol. Manag.* 2007, 244:129-140.
98. Amaranthus, M.P. and Perry, D.A. *Can. J. Forest Res.* 1987, 17:944-950.
99. Byrd, K.B., Parker, V.T., Vogler, D.R. et al. *Can. J. Botany* 2000, 78:149-156.
100. Harvey, A.E., Dumroese, D.S.P., Jurgensen, M.F. et al. *Plant Soil* 1997, 188:107-117.
101. Launonen, T.M., Ashton, D.H. and Keane, P.J. *Plant Soil* 1999, 210:273-283.
102. Mahmood, S., Finlay, R.D. and Erland, S. *New Phytol.* 1999, 142:577-585.
103. Amaranthus, M.P. 1992. *In:* Mycorrhizas in Ecosystem. CAB International, Wallingford, England, 1992, pp. 199-207.
104. Ragazzi, A., Moricca, S., Dellavalle, I. et al. *In:* Decline of Oak Species in Italy; Problems and Perspectives. Ragazzi, A., Dellavalle, I., Moricca, S. et al. (eds.), Università di Firenze, Concilio Nazionale di Ricerca, Accademia Italiana di Scienze Forestali, Firenze, Italy, 2000, pp. 39-75.
105. Ragazzi, A., Vagniluca, S. and Moricca, S. *Physiopathol. Mediterr.* 1995, 34:207-226.

106. Delatour, C. *Rev. Forest Fr.* 1983, 35:265-281.
107. Manion, P.D. *Tree Disease Concepts*, 2nd ed., Prentice Hall, Englewood Cliffs-NJ, p. 402.
108. Schütt P. and Cowling, E.B. *Plant Dis.* 1985, 69:548-558.
109. Manion, P.D. and Lachance, D. *In:* Forest Decline Concepts, an overview. Manion, P.D. and Lachance, D. (eds.), The American Phytopathological Society, St. Paul-MN, 1992, pp. 181-190.
110. Oleksyn, J. and Przbyl, K. *Eur. J. Forest Path.* 1987, 17:321-336.
111. Anselmi, N., Cellerino, G.P., Franceschini, A. et al. *In:* Proceeding of SIPaV/AIPP Congress. L'endofitismo di funghi e batteri patogeni in piante arboree ed arbustive, 19-21 May 2002, Sassari Tempio Pausania, Italy, Franceschini, A. and Marras, F. (eds.), 2002, pp. 6-7.
112. Halmschlager, E. *In:* Proceedings of the Division 7 IUFRO Working Party, Disease/Environmental Interactions in Forest Decline, 6-21 March 1998, Vienna, Austria, Cech, T., Hartmann, G. and Tomiczek, C. (eds.), 1998, pp. 49-56.
113. Marçais, B., Caël, O. and Delatour, C. *Forest Pathol.* 2000, 30:7-17.
114. Cellerino, G.P. and Gennaro, M. *In:* Decline of Oak Species in Italy; Problems and Perspectives. Ragazzi, A., Dellavalle, I., Moricca, S., et al. (eds.), Università di Firenze, Concilio Nazionale di Ricerca, Accademia Italiana di Scienze Forestali, Firenze, Italy, 2000, pp. 157-175.
115. Vannini, A., Valentini, R. and Luisi, N. *Ann. Forest Sci.* 1996, 753-760.
116. Blascke, H. *Eur. J. Forest Path.* 1994, 24:386-398.
117. Causin, R., Montecchio, L. and Mutto Accordi, S. *Ann .Forest. Sci.* 1996, 53:743-752.
118. Gullaumin, J.J., Bernard, C.H., Delatour, C. et al. *Ann. Sci. Forest.* 1985, 42:1-22.
119. Perrin, R. and Estivalet, D. *Agric. Ecosyst. Environ.* 1990, 28:381-387.
120. Thomas, F.M. and Büttner, G. *In:* Proceedings of the International Congress Recent Advances in Studies on Oak Decline. Luisi, N., Lerario, P. and Vannini, A. (eds.), September 13-18, 1992, Selva di Fassano, Brinidisi, Italy, 1992, pp. 285-291.
121. Nagarajan, G. and Natarajan, K. *World J. Microbiol. Biotechnol.* 1999, 15:197-203.
122. Schütt, P. and Cowling, E.B. *Tree Disease* 1985, 69:548-558.
123. Pugh, G.J.F. *Trans. Br. Mycol. Soc.* 1980, 75:1-14.
124. Loureau, M., Naeem, S., Inchausti, P. et al. *Science* 2001, 294:804-808.
125. Zhou, M. and Saharik, T.L. *Can. J. Forest Res.* 1997, 27:1705-1713.
126. Agerer, R. and Waller, K. *Mycorrhiza* 1993, 3:145-154.
127. Hobbie, E.A., Weber, N.S. and Trappe, J.M. *New Phytol.* 2001, 150:601-610.
128. Hobbie, E.A., Weber, N.S., Trappe, J.M. et al. *New Phytol.* 2002, 156:129-136.
129. Hibbet, D.S., Gilbert, L.B. and Donoghue, M.J. *Nature* 2000, 407:506-508.
130. Koide, R.T., Sharda, J.N., Herr, J.R. et al. *New Phytol.* 2008, 178:230-233.
131. Johnson, N.C., Graham, J.H. and Smith, F.A. *New Phytol.* 1997, 135:575-585.
132. Johnson, N.C. *J. Appl. Ecol.* 1998, 35:86-94.
133. van der Heijden, M.G.A., Klironomos, J.N., Moutoglis, P. et al. *Nature* 1998, 365:69-72.
134. van der Heijden, M.G.A., Wiemken, A. and Sanders IR. *New Phytol.* 2003, 157:569-578.

135. Begon, M., Harper, J.L. and Townsend, C.R. Ecologia, Individui, Popolazioni, Comunità. Zanichelli Ed. Bologna, Italy, 1989, pp. 854.
136. Hunt, H.W. and Wall, D.H., *Global Change Biology* 2002, 8:33-50.
137. Fonseca, C.R. and Ganade, G. *J. Ecol.* 2001, 89:118-125.
138. Copley, J. *Nature* 2000, 406:452-454.
139. Shachar-Hill, Y. *New Phytol.* 2007, 174:235-240.
140. St. John, T. *Land and Water* 1998, 42:17-19.
141. Malajczuk, N., Reddel, P. and Brundrett, M. *In:* Mycorrhizae and Plant Health. Pfleger, F.L. and Linderman, R.G. (eds.), PAS Press, St. Paul, Minnesota, 1994, pp. 83-100.
142. Le Tacon, F., Alvarez, I.F., Bouchard, D. et al. *In:* Mycorrhizas in Ecosystem. Read, D.J., Lewis, D.H. and Fitter, A.H. et al. (eds.), CAB International, Wallingford, England, 1992, pp. 199-207.

3

Role of VAM in Nutrient Uptake of Crop Plants

V. Davamani, A.C. Lourduraj[1], R.P. Yogalakshmi and M. Velmurugan[2]

Agro-Climate Research Centre, Tamil Nadu Agricultural University
Coimbatore 641003, Tamil Nadu, India
[1] Water Technology Centre, Tamil Nadu Agricultural University
Coimbatore 641003, Tamil Nadu, India
[2]Department of Vegetable Crops, Tamil Nadu Agricultural University
Coimbatore 641003, Tamil Nadu, India

Introduction

Microorganism and their role in soil must be considered as a key component among those conferring soil fertility and productivity [1]. One of the most important plant microbe interactions is the association of vesicular-arbuscular mycorrhizae (VAM) fungi with higher plants. VAM fungi and rhizosphere microorganism could influence their own mutual development which results in a synergistic interaction. VAM fungi and its interaction with other rhizosphere microorganism improve the plant growth and productivity mainly through increased nutrient assimilation, especially P in nutrient deficient soils.

Agricultural and horticultural crops have been shown to benefit from VAM on a world-wide basis [2]. The potential application of VAM to the agricultural and horticultural industries is great [3]. Since the fungi involved in this symbiotic host root:fungus association can differ in the degree to which they benefit host growth, it is necessary to screen a broad spectrum of VA mycorrhizal fungi (VAMF) to determine the most promising combination for maximum growth response for a particular crop. The relative host growth benefits, especially improved plant phosphorus nutrition, may vary with soil type, host cultivar [4], fungal susceptibility to hyperparasitism [5], and a number of other factors.

Even though there are many reports characterizing specific combinations of mycorrhizal fungal species:crop species, there is a lack of information relating comparative effects of various species on the growth of one or more crops in Florida or elsewhere. Abbott and Robson [6] found that of four isolates of VAMF, two isolates of *Glomus monosporus* and one of *G. mosseae* increased the dry weight and phosphorus content in the tops of subterranean clover (*Trifolium subterraneum* L. cv. 'Seaton Park') over plants inoculated with *G. fasciiilatus* or uninoculated plants. Growth effects of seven different Endogone (*Glomus sensu*) strains on onion (*Allium cepa* L.) and bahiagrass (*Paspalum notatum Fluegge*) varied greatly with the strain used [7]. The tested strains resulted in a 2-15-fold growth increase in onions, and a 2-10-fold increase in bahia grass growth.

Mycorrhizae

VAM association is the most prevalent and important mutualistic association in the plant kingdom. The first person to use the term 'Mycorrhizae' was the German botanist A.B. Frank [8]. He described two types of mycorrhizae: ecto and endo mycorrhizae. Ectomycorrhizae characterized by dense mycelial sheath around the roots and the intercellular hyphal invasion of the root cortex, are limited to mostly temperate forest trees. Endomycorrhizae, where the fungi from external hyphal network in the soil grow extensively within the cells of the cortex, are found nearly in all the plants.

VAM fungi are mainly used in crop plants for better nutrient utilization specially 'P' nutrient and VAM fungi association is wide spread one and most ancient [9]. VAM are not host specific. Nearly 83% of dicotyledens and 79% of monocotyledons form mutualistic symbiosis [10, 11]. Dangeard [12] was the first to name VAM fungus. He described a typical VAM and named the endophyte as *Rhizophagus populins*. VAM have ubiquitous geographical distribution and they occur in plants grown in artic, temperate and tropical regions [13]. Recently Morton and Benny [14] have revised the classification and the VAM forming genera and species have been placed under a new order *Glomales* with three families and six genera:

I	Order	Endogonales
	Family	Endogonaceae
	Genera	Endogone and Sclerogone
II	Order	Glomales
	Sub order I	Glomineae
	Family I	Glomaceae
	Genera	Glomus and Sclerocystis
	Family II	Acaulosporaceae
	Genera	Acculospora and Entrophosphora

Sub order II   Gigasporineae
Family III   Gigasporaceae
Genera   Gigaspora and Scutcillospora

VAM Fungal Association with Various Crops

Mycorrhizal fungi are associated with different type of plants such as onion [15], barley [16], grasses [17], legumes [18], citrus, grapes [19], cassava, olive, pineapple, rubber, papaya, coffee, tea, oil palm, carrot, oats, wheat and marigold [20, 21]. Iqbal et al. [22] reported that young and mature crops of oat, sorghum, barley, wheat and rice were infected with VAM fungi under field condition in Pakistan. Approximately 14 VAM spores of different genera (Gigospora, Acualospora, Entrophophora and Glomus) have been reported in south western Australian agricultural fields [23]. Several reports are also available on the distribution of VAM fungi in Indian soils. Rao et al. [24] reported occurrence of VAM fungi in root system of aromatic plants grown in low 'P' soils. Goje et al. [25] reported that five species of Sclerocystis were found to occur in some tropical deciduous forest species of Andhra Pradesh. Padmavathi et al. [26] also reported widespread occurrence of VAM endophyte in both black and red soils with varying physico-chemical characteristics but could not establish positive correlation between VAM spore density and the soil properties.

Vijayakumar [27] reported that VAM fungi were found to occur in the rhizophere soils of groundnut, sorghum and mulberry grown in three different soils types in Andhra Pradesh and the infection percentage ranged 75-95. Muthukumar and Udaiyan [28] reported sixty species of Pteridophytes, 85 percent had VAM association in their roots and reported the occurrence of *Acaulospora, Glomus, Sclerocystis,* and *Scutellospora* sp. in samples collected from Western Ghats of Southern India.

Among the three VAM fungi viz., *Gigaspora margarita, Glomus fasciculatum* and *Glomus mosseae* compared on the growth of apple seedlings in 'P' deficient soil, *Glomus fasiculatum* was reported to be most efficient [29]. The important factor to be considered in studying effectiveness of a fungus is its ability to take up phosphorus not only in soils with a low nutrient content but also in phosphate amended soil in order to achieve a better use of fertilizers [30]. The VAM inoculation increased the concentration of P, Zn, Cu and Mn in roots and leaf chlorophyll and growth (e.g. leaf area, plant height, dry weight) of the apple plant [31].

VAM Application and Hormonal Changes

The application of plant growth regulators by themselves and in combination with other substances is commonly used to increase adventitious rooting on cuttings. Mycorrhizal fungi are known to produce many plant hormones [32, 33] and polyphenolic compounds which decrease auxin oxidation [34]. Adding mycorrhizal fungi into the rooting medium could result in responses on cuttings similar to that achieved from exogenous application of plant growth regulators, or interact with plant growth regulators in a synergistic or antagonistic fashion.

Chemical Changes Associated with VAM Inoculation

Many changes in metabolism are known to occur during adventitious root formation including changes in amino acids and proteins important for enzyme function and nitrogen metabolism, and changes in carbohydrates [35]. With miniature roses differences in total amino acid, protein, and carbohydrates in cuttings was tracked and comparison of how mixing VAMF into the rooting medium changes composition during the initial stages of rooting. Differences in protein and amino acids between cuttings exposed to inoculum and cuttings made with no inoculum were detectable within two to four days after cutting while differences in carbohydrates were detectable within four to seven days after cutting [32].

VAM and Water Use Efficiency

Mycorrhizal fungi improved water use efficiency in *Rosa hybrida* plants [36] and in safflower and wheat [37]. Mycorrhizal inoculation may directly enhance root water uptake providing adequate water to preserve physiological activity in plants, especially under severe drought conditions [38, 39]. Mycorrhizal inoculation also improves root P uptake, particularly under dry soil conditions [40]. Mycorrhizae, therefore, are likely to be important for increasing P acquisition during drought periods [41].

Mycorrhizal colonization may increase root length density or alter root system morphology, enabling colonized plants to explore more soil volume and extract more water than uncolonized plants during drought [42, 43]. Mycorrhizal hyphae may also directly enhance root water uptake, increasing water supply which would help sustain physiological activity within plants [44]. Mycorrhizal colonization enhances stomatal control in rose plants and reduces water loss during drought [45]. Moreover, improved drought tolerance and better drought recovery by mycorrhizal plants has been linked to improved P uptake [46-48].

Most of the experiments have indicated that VAM was able to alter water relation of its host plants. Mechanism that VAM can enhance resistance of drought stress in host plant may include many possible aspects: (1) VAM improves the properties of soil in Rhizosphere; (2) VAM enlarges root areas of host plants, and improves its efficiency of water absorption; (3) VAM enhances the absorption of P and other nutritional elements, and then improves nutritional status of host plant; (4) VAM activates defense system of host plant quickly;

(5) VAM protects against oxidative damage generated by drought or (6) VAM affects the expression of genetic material [49].

More and more experiments have indicated that VAM was able to alter water relations and played a great role in the growth of host plant in the condition of drought stress [50]. VAM symbiosis improved absorption capacity, and increased the growth of its host plant, which was proved in sugarcane, mung bean, apple, orange, wheat, tomato and wild jujube [51]. There is a great correlation between nutritional status of plant and its drought resistance, while VAM changed the nutritional status of its host plant. P concentrations themselves may affect host water balance, but it is often fixed in soil and not available to plant. Phosphatase produced by VA fungi play an important role in translating fixed or insoluble into soluble P, which can be used by plant freely. At the same time, hyphae are also important ways of P transported in soil. Other elements such as Zn and Cu can also not flow freely in soil [52]. The absorption of Ca, Si, Ni, Co etc. was also reported increased by VAM symbiosis [53].

VAM and Nutrient Uptake

Control of soil microbe is necessary to optimize N and P nutrition of plants. Before attaining the true management of soil microbes, it is essential to understand better the interactions between plants and microbes in the soil around the roots [54]. There are extensive microbial activities in rhizosphere soil which is colonized by a wide range of microbes having important effects on plant nutrition, growth and health. VAM and arbuscular mycorrhizas are mutualistic symbioses formed between the roots of most plants and fungi in the order Glomales [55, 56]. VAM fungi assist the plants to absorb mineral nutrients from the soil, particularly low available elements like phosphorus (P), molybdenum (Mo) and cobalt (Co) [57]. These fungi are also reported to consistently stimulate plant absorption of zinc (Zn) and copper (Cu) and also increase plant resistance to various stresses like water, drought, salt and heavy metal toxicity [57, 58].

For more than two decades now, investigations have increasingly revealed the role of symbionts like VAM fungi. Besides being noted for guaranteeing the availability of plant nutrients (especially phosphates) in soil and subsequent plant roots' absorption of the nutrients [59], they are also known in the biological control of root pathogens, and the improvement of drought tolerance of plants [60] and increasing crop yield [61]. Mycorrhiza is a term used to denote the unique structure formed by the interaction of the roots of higher plants and the fungal mycelium [62]. The VAM is formed by certain fungal species of the family endogonaceae identifiable by their characteristic spores and structures formed on the external hyphae, which can initiate mycorrhiza associations. Mycorrhiza is found throughout the world, and is known to be beneficial to plants. They commonly associate with almost all plant species of agricultural importance. VAM are of the highest spread, and occur on more plant species than any other mycorrhiza [63]. VAM infection enhances the uptake of nutrients (especially phosphorus) by crops [60, 64]. However, Harley and Smith [65] reported that P uptake still depends on the soil phosphorus content.

Application of Azospirillum and VAM along with 50% recommended dose of fertilizers significantly influence the physiological and biochemical behaviour in relation to flowering in mango cv. Himsagar. It is suggested that higher intensity of nitrogen fixing microbial colony improved the nutritional status of plants, which enhance the growth response, flowering and related biochemical parameters [66]. Papaya (*Carica papaya* cv. Coorg Honey Dew) plants inoculated with the VA mycorrhizal fungi *Glomus mossae* and *G. fasciculatum* in sterilized nursery soil showed improved plant height, dry matter as well as P, N and Zn concentrations with no or low levels of phosphorus application. There was an enhanced alkaline and acid phosphatase activity on the root surface and also in the enzyme extract of the root of papaya [67].

VAM and 'N' Nutrition

VAM fungi play a major role in plant N acquisition in many experiments in 'P' deficient soils; nodulation and symbiotic N-fixation were improved on mycorrhizal infection [68, 69]. VAM fungi utilized both the form of NH_4^+ ions [70]. Later Smith et al. [71] showed that VAM increased the 'N' inflow to the plant. Padma [72] reported increase in 'N' content by 69% in CO_3 variety of papaya upon the inoculation of *Glomus fascleulatum*. The N content of the root and leaves of *Citrus sinensis* was increased due to inoculation of *Glomus epigaeum* [73]; *Glomus faciculatum* significantly increase the N uptake in brinjal at all levels of P [74]. Apart from these, VAM fungi were found to enhance the biological N-fixation and thereby facilitating N input into the plant soil system [75]. Inoculation of *Glomus desertieola* induced greater uptake of N by roots of cassava, alley cropped with Glyricida and Leucaena than roots of cassava alley cropped with Senna and sole cassava [76]. Active uptake of NO_3^- by external mycelium of *Glomus intraradices* is monoxenic culture and has been reported by Bago et al. [77]. Azcon et al. [78] reported that the NO_3^- mobility is severely restricted by drought due to its own concentration and diffusion rate. Thus under such condition, the role of mycorrhizae in NO_3^- transport to the root surface may be significant. Significant higher nitrogen concentrations have also been reported in mycorrhiza inoculated plants than in non-mycorrhizal plants, suggesting that VAM hyphae can use N forms that are less available to non-mycorrhizal plants [61].

VAM and 'P' Nutrition

Criteria used in the selection of mycorrhizal fungi for the uptake and transport of 'P' from soil to host have been reviewed by Abbott et al. [79]. Reddy et al.

[80] screened thirteen species of VAM fungi for their symbiosis effect on acid lime in pot experiments and observed that inoculation of *Glomus mosseae, Acaulospora laevis, Glomus caladonium* and *Gigaspora margarita* resulted in greater plant biomass, plant growth, leaf area, leaf number and leaf P and zinc concentration. *G. macrocarpum* and *G. mosseae* were considered to be best in mobilizing 'P' nutrient. Dodd and Thomson [81] reported that the enhanced growth has been attributed to the plants by increasing the uptake of nutrients such as P, Zn, Cu and N. Bolan [82] suggested the different mechanism for the increased 'P' uptake by mycorrhizal plants which includes: (i) physical exploration of soil; (ii) faster movement of 'P' into mycorrhizal hyphae; and (iii) modification of root environment. It has been shown that the external hyphae of VAM fungi could deliver up to 80% of the plant 'P' to the host plant over a distance of more than 10 cm from root surface [83, 84].

Thiagarajan and Ahmed [85] examined the effect of a VAM fungus, *Glomus palldam* on phosphates activity in cowpea roots at successive stages of plant growth. Both acid and alkaline phosphates activities were significantly higher in mycorrhizal than non-mycorrhizal roots. Smith [86] concluded that mycorrhiza fungi directly increased the rate of phosphate uptake by roots over a range of soil phosphorus level, even when mycorrhiza growth has seized. Kumar et al. [87] used alkaline phosphates level as indicator for screening and differentiating efficient VAM fungi in spring wheat and reported that *Glomus versiforme* showed higher alkaline phosphatase activity as compared to other isolates tested. Kojima et al. [88] confirmed the presence of alkaline phosphatases in intraradical hyphae of arbuscular mycorrhizal fungi by isozyme studies and it was found to be closely related to the improvement of plant growth. Kennady and Rangarajan [89] studied the effect of six different VAM fungi in three papaya varieties for increase of biomass and phosphatase activity. Acid phosphatase activity by VAM colonization enhanced significantly in all three varieties than alkaline phosphatase activity.

Conclusion

Due to the increasing of input cost especially fertilizer cost, the cultivation of crop area is reduced day to day. Apart from this, continuous use of fertilizer cause environmental contamination and pollution hazards increase. Now, we are in a position to reduce the use of inorganic fertilizer and increase the use efficiency of applied nutrients. Possibilities for the improvement of nutrient use efficiency include use of VAM fungi and use of chemical growth substances. The mechanism for the activation of fixed nutrients (particularly P and K) by mycorrhizae fungi needs to be carried out in future and crop responses to the interaction of VAM and organic nutrients in acid soils and uptake capacity of various major nutrients should also be evaluated.

References

1. Pauli, F.W. A bio-dynamical approach. Adam, Hilger, London, 1967.
2. Gerdemann, J.W. *Ann Rev Phytopathol* 1968, 6:397-418.
3. Ruehle, J.L. and Marx, D.H. *Science* 1979, 206:419-422.
4. Menge, J.A., Johnson, E.L. and Platt, R.G. *New Phytol* 1978, 81:553-559.
5. Daniels, B.A. and Menge, J.A. *Phytopathology* 1980, 70:584-588.
6. Abbott, L.K. and Robson, A.D. *Anst J Agric Res* 1977, 28:639-649.
7. Mosse, B. *Nature* 1969, 239:221-223.
8. Frank, A.B. *Ber Dtsch Bot Gen* 1885, 3:128.
9. Simon, I., Bousquert, J., Levesque, R.C. et al. *Nature* 1993, 363:67-69.
10. Bonfante-Fasolo, P. *Symbiosis* 1987, 3:249-254.
11. Trappe, J.M. *In:* The angiosperms form an evolutionary stand point, Ecophysiology of VAM plants. Safir, G.R. (ed.), CRC Press, Boca Raton, Florida, 1987.
12. Dangeard, P.A. *In:* Progress in microbial ecology. Mukherji, K.G., Agnihorti, V.P. and Singh, R.P. (eds.), Print House, Lucknow, 1900, pp. 489-524.
13. Mosse, B., Stribly, D.P. and Le Tacon, F. *Adv Microb Ecol* 1981, 5:137-210.
14. Morton, J.B. and Benny, G.L. *Mycotaxon* 1990, 37:471-491.
15. Gerdemann, J.W. *Ann Rev Phytopathol* 1968, 6:397-418.
16. Saif, S.R. and Khan, A.G. *Plant Soil* 1977, 47:17-26.
17. Nicolson, T.N. *Trans Brit Mycol Soc* 1959, 42:421-438.
18. Butler, E.J. *Trans Brit Mycol Soc* 1939, 22:274-301.
19. Mosse, B. *Nature* 1957, 179:922-924.
20. Hayman, D.S. *In:* Advances in Agricultural Microbiology. Subba Rao, N.S. (ed.), Oxford and IBH Publishing Co, New Delhi, 1982.
21. Plenchette, C., Fortin, J.H. and Furlan, V. *Plant Soil* 1983, 70:199-209.
22. Iqbal, S.H., Taquir, T. and Ahmad, J.H. *Plant Soil* 1978, 63:505-509.
23. Hall, I.R. and Abbott, L.K. *Trans Brit Mycol Soc* 1984, 83:203-208.
24. Rao, Y.S.G., Suresh, C.K., Suresh, N.S. et al. *Curr Res* 1987, 16:55-57.
25. Goje, L.R., Ramapulla Reddy, P., Sathya Prasad, K. et al. *Indian J Microbiol Ecol* 1991, 2:127-131.
26. Padmavathi, T., Veeraswamy, J. and Venkateswarlu, K. *J Indian Bot Soc* 1991, 70:193-195.
27. Vijayakumar, R.S. *J Hill Res* 1998, 11:8-11.
28. Muthukumar, T. and Udaiyan, K. *Phytomorphology* 2000, 50:132-142.
29. Geddida, Y.I., Trappe, J.M. and Stebbins, R.L. *J Am Soc Hort Sci* 1984, 109:24-27.
30. Schubert, A. and Hayman, D.S. *New Phytol* 1986, 103:79-90.
31. Runjin, Xinshu. *In:* International Symposium on Diagnosis of Nutritional Status of Deciduous Fruit Orchards. Indian Society Horticulture Science, Acta Horticulturae, 1999, p. 274.
32. Scagel, C.F. and Linderman, R.G. *Tree Physiol* 1998, 18:739-747.
33. Scagel, C.F. and Linderman, R.G. *Symbiosis* 1998, 24:13-34.
34. Mitchell, R.J., Garrett, H.E., Cox, G.S. et al. *Tree Physiol* 1986, 1:1-8.
35. Hassig, B.E. *In:* New Root Formation in Plants and Cuttings. Jackson MB (ed.), Martinus Nijhoff Pub, Dordrecht, 1986, pp. 141-189.
36. Henderson, J.C. and Davies, F.T. *New Phytol* 1990, 115:503-510.

37. Bryla, D.R. and Duniway, J.M. *Plant Soil* 1997, 197:95-103.
38. Faber, B.A., Zasoski, R.J., Munns, D.N. et al. *Can J Bot* 1991, 69:87-94.
39. Smith, S.E. and Read, D.J. Mycorrhizal Symbiosis, 2nd ed. Academic Press, New York, 1997.
40. Jakobsen, I. *In:* Mycorrhiza: Structure, Function, Molecular Biology and Biotechnology. Varma, A. and Hock, B. (eds.), Springer-Verlag, Berlin, 1995, pp. 297-324.
41. Bryla, D.R. and Duniway, J.M. *New Phytol* 1997, 136:581-590.
42. Kothari, S.K., Marschner, H. and George, E. *New Phytol* 1990, 116:303-311.
43. Davies, F.T., Svenson, S.E., Henderson, J.C. et al. *Tree Physiol* 1996, 16:985-993.
44. Allen, M.F. *New Phytol* 1982, 91:191-196.
45. Auge, R.M., Schekel, K.A. and Wample, R.L. *New Phytol* 1986, 103:107-116.
46. Nelson, C.E. and Safir, G.R. *Planta* 1982, 154:407-413.
47. Graham, J.H., Syvertsten, J.P. and Smith, M.L. *New Phytol* 1987, 105:411-419.
48. Fitter, A.H. *J Exp Bot* 1988, 39:595-603.
49. Song, H. *Electronic Journal of Biology* 2005, 1(3):44-48.
50. Augé, R.M. *Mycorrhiza* 2001, 11:3-42.
51. Wu, Q. and Xia, R. *Chinese Agricultural Science Bulletin* 2004, 20:188-192.
52. Li, X., Marschner, H. and George, H. *Plant and Soil* 1991, 136:49-57.
53. Gong, Q., Xu, D. and Zhong, C. Chinese Forest Press, Beijing, China, 2000, 51-61.
54. Tacon, F., Le, R.D., Fraga-Beddiar, A. et al. *In:* Research on Multipurpose Tree Species in Asia. David, A., Taylor, K. and Mac Dicken, G. (eds.), Winrock International Institute for Agricultural Development, 1971.
55. Menge, J.A. *Can J Bot* 1983, 61:1015-1024.
56. Tisdale, S.L., Nelson, W.L. and Baton, J.D. *World J Agri Sci* 1995, 2(1):16-20.
57. Mosse, B. *Ann Rev Phytopath* 1973, 11:171-196.
58. Jackson, R.M. and Mason, P.A. Mycorrhiza. Edward Arnolds Ltd., London, 1984, p. 60.
59. Sieverding, E. *Agri Ecosys Environ* 1981, 29:369-390.
60. Osinubi, O., Bakare, O.N. and Mulongoy, K. *Bio Fert Soils* 1992, 14:159-165.
61. Taiwo, L.B. and Adegbite, A.A. *Moor J Agri Res* 2001, 2:110-118.
62. Tate, R.L. Soil Microbiology. John Wiley and Sons Inc., New York, 1995.
63. Werner, D. Symbiosis of plants and microbes. Chapmen and Hall, London, 1992.
64. Atayese, M.O. and Laisu, M.O. *Moor J Agri Res* 2001, 2:103-109.
65. Harley, J.L. and Smith, S.E. Mycorrhizal Symbiosis. Academic Press, London, 1983.
66. Das, K.K., Mandal, M.A., Hasan, B. et al. *In:* VIII International Mango Symposium, Indian Society Horticulture Science, Acta Horticulturae, 2000, p. 820.
67. Mohandas, S. *Nutrient Cycling in Agroecosystems* 1992, 31(3):263-267.
68. Crush, J.R. *New Phytol* 1974, 73:743.
69. Smith, S.E. and Daft, M.J. *Aust J Plant Physiol* 1977, 4:403-413.
70. Bowen, G.D. and Smith, S.E. *Ecological Bulletin* 1981, 33:109-116.
71. Smith, S.E., St John, B.J., Smith, F.A. et al. *New Phytol* 1986, 103:359-373.
72. Padma, T.M.R. MSc Thesis, Tamil Nadu Agricultural University, Coimbatore, Tamil Nadu, India, 1988.

73. Chang, B.K. and Chien, K.S. *Agri Ecos Environ* 1989, 29:35-38.
74. Merina, P.S. MSc Thesis, Tamil Nadu Agricultural University, Coimbatore, Tamil Nadu, India, 1991.
75. Onkarayya, H. and Mohandas, S. *Adv Hort Forestry* 1993, 3:81-91.
76. Osonubi, O., Atayese, M.O. and Mulongoy, K. *Biol Fertil Soils* 1995, 20:70-76.
77. Bago, B., Vierheilig, H., Piche, Y. et al. *New Phytol* 1996, 133:273-280.
78. Azcon, R., Gomez, M. and Tobar, R. *Biol Fertil Soil* 1996, 22:156-161.
79. Abbott, L.K., Robson, A.D. and Gazey, C. *In:* Methods in microbiology, Techniques for study of mycorrhiza. Norris, J.R., Reed, D.J. and Karma, A.V. (eds.), Academic Press, London, 1992, 24:1-22.
80. Reddy, B., Bagyaraj, D.J. and Mallesha, B.C. *Indian J Microbiol* 1996, 36:13-16.
81. Dodd, J.C., Thomson, B.D. *Plant Soil* 1994, 159:149-158.
82. Bolan, N.S. *Plant Soil* 1991, 134:189-207.
83. Rhodes, L.H. and Gardemann, J.W. *New Phytol* 1975, 75:555-561.
84. Li, X.L., Marschner, H. and George, E. *Plant Soil* 1991, 136:49-57.
85. Thiyagarajan, T.R. and Ahmed, M.H. *Boil Fertil Soils* 1994, 17:51-56.
86. Smith, A.N. *Plant Soil* 1965, 23:314-316.
87. Kumar, P.C., Garibova, C.L.V. and Velikanov, L.L. *Microbiology and Biotechnology*, 1999, pp. 67-69.
88. Kojima, T., Hayatsu, M. and Saito, M. *Biol Fertil Soils* 1998, 26:335-337.
89. Kennady, Z.J. and Rangarajan, M. *Indian Phytopathol* 2000, 53:38-43.

4

Advances in Mass Production Technology of Arbuscular Mycorrhiza

A. Khaliq, D.J. Bagyaraj[1] and M. Alam

Central Institute of Medicinal and Aromatic Plants (CIMAP)
P.O. CIMAP, Lucknow 226015, India

[1]Centre for Natural Biological Resources and Community Development
41 RBI Colony, Anand Nagar, Bangalore 560024, India

Introduction

The term mycorrhiza, derived from two words 'mycos' and 'rhiza', literally means 'fungus root'. Frank (1885) for the first time proposed this term to describe the mutualistic association occurring between plant root and fungi [1]. Mycorrhizal associations are categorized into two classes: Ectomycorrhiza and Endomycorrhiza. The ectomycorrhiza are, generally, associated with the roots of higher woody perennials. These fungi form intercellular hyphal ramification with the host cortex i.e., hartig net and dense hyphal covering to roots (i.e., mantle). Endomycorrhiza is characterized by fungal penetration of the host cell where three categories are recognized. These are orchidaceous, ericoid and arbuscular. The associated endophyte in first two belongs to higher fungi with septate hyphae. An intermediate group with septate, inter- and intracellular mycelium was placed in a separate group called Ectendomycorrhiza [2]. Arbuscular mycorrhizal (AM) symbiosis, also referred as vesicular-arbuscular mycorrhizal symbiosis [3], is the most common and has been estimated to occur in more than 80% of flowering plant species [2]. These associations are extremely ancient and AM fungi have even been identified in fossils of early Devonian land plants [4].

The AM fungi are obligatory biotrophic in nature, colonizes the cells of the roots in order to obtain nutrition from the host plant. In addition to growth within the roots by forming intracellular dichotomously branched structures, the arbuscules, which are the site for translocation of minerals between plant and roots, they form a network of external hyphae which absorbs phosphates and other mineral nutrients from the soil and translocate to the root. Improvement in plant growth and biomass, production of healthy plants, soil stabilization and development of resistance against soil-borne plant pathogens due to AM infection reveal that these fungi play an important role in the ecosystem [5-13]. They make association with the plants found mostly in all soils and weather conditions like desert [14], aquatic conditions [15], salt marsh [16] and highly alkaline usar soil [17].

Progress in the development of inoculum of AM fungi has been severely hampered by inability to grow them in axenic culture. This might be due to the obligate biotrophic nature of the fungus; therefore, they must be grown in the presence of living host root system. The potential utilization of AM fungi highly depends on a suitable inoculum that can be produced in a simple way and be spread on agricultural land with traditional equipments. Several techniques of inoculum production have been developed time to time by different mycorrhizologists for the sake of its practical utilization and understanding its physiological, biochemical and genetic characteristics, which are discussed in this review.

Classical Methods

Soil-Root Inoculum

This is the standard and conventional way of maintaining AM culture around the world. The most common medium used so far for this purpose is sterilized soil, which is inoculated with pure stock culture of a particular AM isolate in a living root system [18, 19]. The inoculum so produced consists of a mixture of soil, spores, pieces of hyphae and infected root pieces and its production generally takes 3-4 months. Although this practice is most common and widely used by the workers, it has several drawbacks. Firstly, very limited amount of inoculum can be produced by this method. Secondly, the product is so bulky and heavy to maintain and transport them to the site of its application. Gradually, inert substrates such as vermiculite, perlite, sand, or a mixture of these replaced the soil. Thirdly, there is a risk of cross contamination of adjacent pots of endophyte, impurity of inoculum containing other microorganisms and lack of genetic stability under these conditions. Existence of different physiological strains within a single morphological species has been reported [20] but such strain differences are difficult to accommodate in a culture collection based on morphological criteria. Similarly, pot culture may include selection of some strains favoured by the particular growth conditions.

Inoculum Rich Soil-Pellets

Hall and Kelson introduced a new technique for AM inoculation [21]. They prepared soil pellets enriched with the inoculum. The pellets measured 12×12×6 mm and had an average dry weight of 1.55 g. These dry pellets are glued with seeds by gum arabic and then used. Using this method, two persons can comfortably make 10,000 pellets in a day, as claimed by the workers. Most time consuming step in that manufacture was ensuring that the clay-binding agent was evenly distributed throughout the soil inoculum. The pellets can easily be broadcasted like other fertilizers and may be spread during seed sowing/transplantation.

Soil-Free Inoculum

Hewitt introduced a new type of inoculum production technique where author used calcined montmorillonite clay as supporting medium with supplementation of Long Ashton mineral solution [22]. It proved to be a good substrate for plant growth and abundant mycorrhizal infection and eliminated the need for soil in the growth medium. Dehne and Backhaus used light expanded clay material as a carrier material in a hydroponic culture system [23]. Vestburg and Sukainen described a new method for producing inoculum in a soil-free substrate [24]. They used polymeric hydrogel product 'Water-Works' (USA) as a soil-free substrate for producing AM inoculum. Vermiculite was used as a co-substrate together with the dehydrated polymer crystals. Bone meal was used as a light fertilizer. This technique had a potential for the production of contamination free inoculum.

Nutrient Film Culture

In this technique of inoculum production, the plant roots lie in a shallow layer of floating nutrient solution [25]. In deep flowing solution culture, sophisticated equipments have been designed for continuous adjustment of pH and nutrient concentration. The inoculum produced by this method provides easily harvestable solid mat of roots with more concentrated and bulky form of inoculum than that produced by plants grown in soil or other solid media. Mosse and Thompson carefully manipulated the nutrient concentration in this culture and used *Rhizobium* nodules for nitrogen supplementation in medium [26]. They obtained upto 60% of mycorrhized roots of beans with attached mycelium, spores, and sporocarps.

Aeroponic Culture

This culture was for the first time introduced by Zobel et al. for the study of legume-rhizobia interaction [27]. The technique involves the production of inoculated roots in a chamber providing nutrient solution in the form of mist. Lack of physical substrate in aeroponic culture makes it ideal system for obtaining sufficient amount of clear propagules, useful not only for inoculation, but also useful for critical physiological and genetic studies of this obligate biotroph. Major advantages of this system over others are lack of physical substrate, control of cultural conditions and sampling of mycorrhiza and associated nutrient solution. Hung and Sylvia [28] refined the technique of Zobel et al. [27] and used bahia grass (*Paspalum notatum*) and sweet potato (*Ipomoea batatas*) and colonized these roots with *Glomus deserticola, G. etunicatum*, and *G. intraradices*. They get abundant arbuscules and vesicles formed in the roots. They also found that aeroponically produced *G. deserticola* and *G. etunicatum* inocula retained their infectivity after cold storage (4°C) in either sterile water or moist vermiculite for at least four and nine months, respectively.

Root Organ Culture

It is really a challenging goal for modern plant biologists to culture the AM fungi under axenic conditions. Mosse in 1962, for the first time, established the culture by inoculating the germinating resting spores of *Endogone* sp. [29]. She demonstrated the necessity of *Pseudomonas* sp. for germination, appresorium formation and even penetration of *Endogone* sp. into the root. Later Mosse and Hepper [30] and Miller-Wideman and Watrud [31] opened the path of *in vitro* culture when they separately used growing roots *in vitro* and succeeded in establishing vesicular-arbuscular mycorrhizal symbiosis. Isolated root can be propagated continuously in different solid and liquid media with high reproducibility. Clonal roots of some 15 plants have been established and this list has enlarged during the last decades. Initiation of isolated root requires pre-germination of seeds previously surface sterilized with classical disinfectants (sodium hypochlorite, hydrogen peroxide), and then thoroughly washed in sterile distilled water. Germination of seeds occurs after 48 hr at 28°C in the dark on water agar or moistened filter papers. The tips (2 cm) of emerged roots can be transferred to a rich medium such as modified White medium [32] or Strullu and Romand medium [33]. The pH of the medium is adjusted to 5.5 before autoclaving. Fast-growing roots are cloned by repeated subcultures.

Culture Establishment on Transformed Roots

Nowadays, the use of genetically transformed roots by Ri plasmid of *Agrobacterium rhizogenes* has given a new impetus to the culture production of these obligate symbionts. The *A. rhizogenes,* a soil dwelling, gram negative bacterium produces a neoplastic plant disease syndrome known as 'Hairy Root' which has ability to grow rapidly showing two-fold multiplication within 45-

48 hr in *Atropa belladona* and *Nicotiana tabaccum* [34]. They are also highly stable and capable of secondary metabolite production. This characteristic is due to presence of integrated copies of transfer DNA (T-DNA) which occurs in Ri (root inducing) plasmid, which is a large plasmid [35]. The T_R-DNA locus of T-DNA of Ri plasmid also carry a gene responsible for the production of certain amino-acid derivatives called opines like agropine, mannopine, cuccumopine and mikimopine, which is indicative of transformation. Similarly loci encoding for auxin and other loci responsible for the production of roots are still the areas of further investigations [36]. The rapid and stable root growth due to modification by *A. rhizogenes* infection is very important and favourable for the mass production of arbuscular-mycorrhizal culture.

Production of Transformed Roots

Hairy roots produced by *A. rhizogenes* infection can be sub-cultured as excised roots (Figure 1). On the solid culture media, they are repeatedly treated with suitable antibiotics like carbenicillin or cephalexin to make the roots free from original bacterial inoculum and thereby establish the isolation of a single root piece clonal culture. An example of the production of such culture from tomato stem tissues [37] is described (Figure 2). Stem segments of 25-30-day-old tomato plant, after surface sterilization with sodium hypochlorite (2% available chlorine) for 5 min., were inoculated with freshly grown bacterium (48 hrs. old in nutrient broth) at distal surface, because this bacterium has a higher

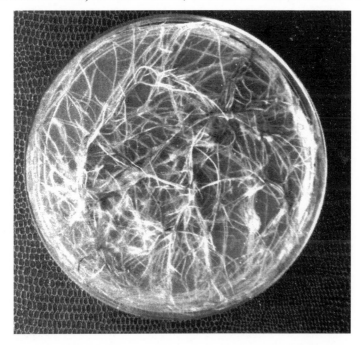

Figure 1: *Agrobacterium rhizogenes* induced clonal culture of tomato roots.

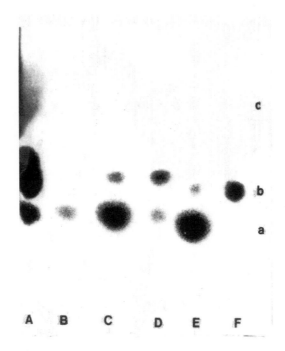

Figure 2: Electrophoretic analysis of opines in *Agrobacterium rhizogenes* mediated transformed roots of tomato. Lane A and F, standards [(a) agropinic acid; (b) mannopine + mannopinic acid; and (c) agropine]; Lane B, non-transformed roots; and Lanes C, D and E are transformed roots incited by *A. rhizogenes* strain ATCC 15834, A4 and LBA 9402, respectively [37].

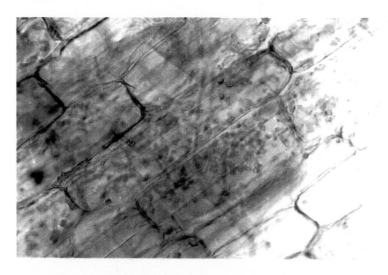

Figure 3: Arbuscules and mycelia produced in the cortical tissues of *Agrobacterium rhizogenes* mediated transformed roots of tomato after inoculation of *Gigaspora margarita* (× 400) [37].

virulence on the distal surface than apical part due to the higher endogenous auxin level [38]. The treated stem segments were placed in Murashige and Skoog (MS) medium and incubated in dark for two days at 25°C. Thereafter, the segments were transferred to MS medium amended with cephalexin (500 mgL^{-1}) and again incubated under the same conditions. After 8-10 days, few transformed roots were found to be proliferated which were later excised and maintained as a clonal culture. Attempts have also been made for successful production of hairy root culture from different plant parts like leaf, cotyledon, anther etc. Successful transformation was later confirmed either by the detection of opines in hairy root tissues [39] (Figure 3) or by molecular hybridization [35].

Establishment of Infection by Mycorrhizal Propagules on Transformed Roots

Selection of Mycorrhizal Propagules

Selection of the type of mycorrhizal propagules is very important for infecting hairy roots. The inoculum should be healthy, viable and contamination-free. Different workers have used different types of inocula. Mosse and Hepper used sporocarp of *Glomus mosseae* [30], Strullu and Romand [33] as well as Zhipeng and Shiuchien [40] used infected root segments colonised with mycorrhiza. Becard and Fortin [32] and Abdul-Khaliq and Bagyaraj [37] used azygospores of *Gigaspora margarita*. Mosse and Hepper [30] stated that since sporocarps are big in size and contain several chlamydospores, they have plenty of chance to take infection in comparison to single chlamydospore. The isolation of chlamydospores generally involves wet sieving and decanting [41], followed by density gradient centrifugation in order to further purify the spores [42]. The latter technique removes dead spores and other soil debris, which is the main source of contamination. While using this technique, Abdul-Khaliq and Bagyaraj [37] mixed a light orange stain (edible) in the middle gradient of sucrose in order to make clear visual separation of different gradients prior to and after the centrifugation.

Surface Sterilization of Mycorrhizal Propagules

Surface sterilization of AM fungal spores is a very important step for establishing infection on the host. A thorough perusal of literature reveals that different workers have used different techniques of spore sterilization [29, 43-47]. Most of them have used chloramine-T (2%), gentamycin and streptomycin sulphate. Abdul-Khaliq and Bagyaraj [37] modified the procedure of sterilization of Mertz et al. [43], which was simple, easy and highly effective than the processes described by earlier workers. In the process, the spores were treated with 2% chloramine-T containing 0.05% tween-20 in an injection vial under light vacuum developed by taking out air from the solution by injection syringe [43]. Light vacuum was able to remove liquefied gases and air droplets adhered to the spore surface. Thereafter, the spores were removed on sterile Whatmann no. 1 filter paper placed in a funnel fitted with rubber tube and stop cock. The spores were washed thoroughly with sterile distilled water to remove tween-20 and chloramine-T. The spores were then treated with sterile antibiotic solution (gentamycin 100 µg mL^{-1} + streptomycin sulphate 200 µg mL^{-1}) for 20 min. They were either utilized immediately or were stored at 4°C for 3 to 4 months in a sealed Petri plate. If the spores were stored for few months, they needed treatment of chloramine-T and antibiotic solution prior to inoculation on hairy roots.

Inoculation of Mycorrhizal Propagules

Solid nutrient media are good substrate for the establishment of dual culture of root and fungus. Mosse and Hepper [30] for the first time established the root organ culture of AM. They divided the medium into two parts, one part was lacking sucrose, and other was a complete medium. They also established the dual culture on liquid medium. Later, Mugnier and Mosse [46] established the root organ culture of Ri T-DNA transformed roots of *Convolvulus sepium* and established the mycorrhizal infection in these roots. They also used bi-compartmental culture system. Roots growing from modified Murashige and Skoog (MS) nutrient medium and sugar into a compartment containing water agar and neutralized peat were infected with germinated spores of *Glomus mosseae*. They suggest the importance of concentration of N in the medium. Accordingly, above 2 mM total N, germ tube of *G. mosseae* stopped growing. In range of 1 to 2 mM N, germ tubes developed but passed across root without becoming attached to the surface. When the nutrient medium contained less than 0.2 mM N, the fungus showed some directional growth towards the root, and the hyphae became attached. They began to form the characteristic, strongly branched, septate fan-like structure on the root surface, three days after inoculation. This was followed by formation of appresorium and root penetration by day 5. The fungus spread intercellularly and arbuscules began to form by day 10. Based on their experiment, they suggested lesser sugar (less than 20 g L^{-1}), essentiality of EDTA and acidic range of pH (below 6) for better germination of spore and penetration of mycelium to the roots.

Becard and Fortin [32], later, improved the methodology and adopted it for the study of initial events of AM ontogenesis. They directly inoculated a single ungerminated spore of *Gigaspora margarita* on a single root system. Since the germ tube growth is negatively geotropic in nature, inserted a single spore into the medium at the bottom of Petri plate, so that germ tube could grow towards the surface of the medium and come in contact with properly placed roots [48]. Thereafter, the process of inoculation was simplified, understanding the negatively geotropic nature of the germ tube. A single spore was placed near the single transformed root in the same Petri plate and incubated the whole Petri plate vertically in such a way that germ tube grow upward and immediately come in contact with the root [37, 49].

Nutrient Medium

The most important factor for successful arbuscular mycorrhiza formation in root organ culture is the adjustment of appropriate culture medium for dual cultivation of both the components. Since the root, which needs rich nutrient medium for its growth and the AM fungi grows normally in relatively poor nutrient conditions, there must be such a balance in the nutrient medium that the growth of the host as well as fungal symbionts should progress during dual culture establishment, keeping in mind the delicacy of obligate biotrophic nature of the AM propagules. The most frequently used media for plant tissue culture are White's [50] and Murashige and Skoog (MS) medium [51]. They are also generally used by different workers for co-cultivating roots and AM propagules. Miller-Wideman and Watrud [31] used 1/10 diluted MS medium for co-culturing *G. margarita* on tomato seedling explant. Although, they had success in establishing the symbiosis, they felt difficulty in root growth evaluation due to poor nutritional status of culture medium. It is more appropriate to use White's medium for root initiation, since this medium has been developed only for root organ culture. Moreover, presence of ammonium ions (in MS medium) was found to be detrimental to root growth because of steep fall in pH of the medium due to this additive. On the contrary, in White's medium, the additive nitrate counteracts the acidification in the culture medium following root growth that buffers the culture medium and retains pH at 6 for several months. Likewise, establishment of mycorrhizae greatly depends upon presence/absence and concentration of sodium sulphate, P, and sucrose in the culture medium. Mosse and Philips [52] reported detrimental effect of sodium on AM establishment as they observed internal development of *Endogone mosseae* in the roots of *Trifolium parviflorum* decrease when sodium was present in the medium. High P level in soil has also been shown to decrease or eliminate mycorrhizal infection [53]. Reducing the concentration of sucrose in the medium also benefits the mycorrhizal colony and arbuscules development, although the root diameter is reduced. The media used by Mosse [54], although, rich in sucrose and P, provided nutritional conditions to stimulate "independent" growth of *G. intraradices* but simultaneously suppressed mycorrhizal infection of trans-formed carrot roots. In this direction the medium modified by Becad and Fortin [32], Mosse and Hepper [30] and by Abdul-Khaliq and Bagyaraj [37] with less in sucrose and P, with no or least ammonium salts, sometime also replacing ammonium salts with nitrate salts, was found suitable for hairy root culture preparation. It has also been convenient for dual culture of *G. intraradices* with transformed roots and/or normal roots of carrot [55] as well as for *G. margarita* with transformed roots of tomato [37]. St. Arnaud et al. [56] grew *G.*

intraradices in transformed roots of modified White's medium [32] in a double compartment system of Petri plate. The upper half compartment had sucrose only where transformed mycorrhizal root system was growing. Only the endosymbiont was permitted to the second (lower) compartment containing the same medium devoid of sucrose. Colonization of the lower compartment by the mycelium took place between six and eight weeks after sub-culturing the mycorrhizal roots in the upper compartment. They counted up to 34,000 spores with a mean of 15,000 mostly viable spores per Petri plate in the distal compartment.

There has been an interest for the production of aseptic spores of AM fungus, mainly for research purpose for almost two decades. In 1984, the production of 3-5 newly formed *G. margarita* spores from Petri plate of tomato root culture was reported [31]. Four years later, modifying the host and growth parameter, Becard and Fortin [32] reported 100 spores per Petri plate of the same AM species and this number was further increased to 450 [57]. Later Diop et al. [58], while working with *G. intraradices* and *G. versiforme* associated with excised tomato roots, repeated the production of 100-1000 aseptic mature spores per Petri plate. Declerck et al. [59] reported production of an average of 9500 spores of *G. versiforme* per Petri plate during transformed carrot root co-culturing in modified Strullu-Romand (MSR) medium [57]. During the same year St. Arnaud et al. [56] made a record of 34,000 spores' production per Petri plate.

Monoxenic culture technology is a very powerful tool for the experimental establishment of arbuscular mycorrhizal associations. A small but growing number of papers have reported the successful establishment of AM associations using a number of AM fungal species [60]. Bi et al. [61] reported for the first time the establishment of an arbuscular mycorrhizal association between *G. sinuosum* (= *Sclerocystis sinuosa*) and transformed Ri T-DNA carrot (*Daucus carota* L.) roots in monoxenic culture. The *G. sinuosum* sporocarps survived not as single spores, but as sporocarps in the environment. The experiment has added *G. sinuosum* to the list of fungi by using monoxenic culture with transformed Ri T-DNA carrot roots.

Root Exudates and Flavonoids

Role of plant exudates and flavonoids is also very important in stimulating the germination as well as growth of AM fungi in dual culture system. It was shown that white clover root-exudate stimulates hyphal elongation of *G. fasciculatum* [45]. Similarly, Gianinazzi-Pearson et al. [62] reported that three flavonoids—apigenin, hesperitin and naringerin—promoted spore germination and hyphal growth of *G. margarita* at concentrations of 0.15-1.5 μM. Quercetin was also found as a highly stimulatory flavonoid which had a synergistic effect with carbon dioxide. Hyphal growth of *G. margarita* was greatly stimulated by a synergistic interaction between volatile and exudated factors produced by the roots. In this connection, Becard and Piche [49] demonstrated carbon dioxide as a critical volatile involved in the enhancement of hyphal growth.

Conclusion

The mass culture production with the help of a bioreactor has given a new dimension for research. Nuutila et al. [47] used small-scale bioreactor for growing *G. fistulosum* with strawberry hairy roots. They obtained up to 3.07 gL^{-1} of dry weight of inoculum. Recently, the production of *G. intraradices* on transformed carrot roots in Petri plate as well as airlift bioreactor has been demonstrated. The results revealed that maximum spore production in the air lift bioreactor was ten times lower than that of Petri plate culture obtained with the lowest inoculum assessed (0.13 g dry wt. L^{-1} medium) with 1.82×10^5 +/– 4.05×10^4 spores (g dry wt. L^{-1} medium)$^{-1}$ in 107 days. Although good growth was obtained in this bioreactor, other factors like root exudates, flavonoids, EDTA, and nutrient minerals should be tested individually for every specific type of inoculum production, especially if the root culture is to be used in large amount in plant root-fungus studies.

References

1. Frank, A.B. *Pilze. Ber. Dt. Bot. Ges.* 1885, 3:128-145.
2. Harley, J.L. and Smith, S.E. Mycorrhizal Symbiosis, Academy Press, London, 1983, p. 483.
3. Walker, C. *In:* Mycorrhiza: Structure, Function, Molecular Biology and Biotechnology. Verma, A. and Hock, B. (eds), Springer, 1995, pp. 25-29.
4. Remy, W., Taylor, T.M., Hass, H. et al. *Proc. Natl. Acad. Sci. USA* 1994, 91:11841-11843.
5. Abdul-Khaliq, Gupta, M.L. and Kumar, S. *Indian Phytopath.* 2000, 54:22-24.
6. Gupta, M.L., Abdul-Khaliq, Pandey, R. et al. *J. Herbs, Spices, Med. Pl.* 2000, 7:57-63.
7. Abdul-Khaliq and Janardhanan, K.K. *Symbiosis* 1994, 16:75-82.
8. Abdul-Khaliq and Janardhanan, K.K. *J. Med. Arom. Pl. Sci.* 1994, 19:7-10.
9. Gupta, M.L. and Janardhanan, K.K. *Plant Soil* 1991, 131:261-263.
10. Koske, R.E. and Halvorson, W.L. *Can. J. Bot.* 1981, 59:1413-1422.
11. Pandey, R., Gupta, M.L., Singh, H.B. and Kumar, S. *Bioresource Technol.* 1999, 69:275-278.
12. Ratti, N., Alam, M., Sharma, S. and Janardhanan, K.K. *Symbiosis* 1998, 24:115-124.
13. Ratti, N., Abdul-Khaliq, Shukla P.K. et al. (unpublished data).
14. Rao, K.B.A.V. and Tarafdar, J.C. *Arid Soil Res. Rehab.* 1989, 3:391-396.
15. Bagyaraj, D.J., Manjunath, A. and Patil, R.B. *Trans. Br. Mycol. Soc.* 1979, 71:164-167.

16. Sen Gupta, A. and Chaudhary, S. *Plant Soil* 1990, 122:111-113.
17. Janardhanan, K.K., Abdul-Khaliq, Naushin, F. et al. *Curr. Sci.* 1994, 67:465-469.
18. Gupta, M.L., Janardhanan, K.K., Chatopadhyay, A. et al. *Mycol. Res.* 1990, 94:561-563.
19. Abdul-Khaliq. Ph.D. Thesis, Dr. Ram Manohar Lohia University, Faizabad, India, 1994.
20. Gildon, A. and Tinker, P.B. *New Phytol.* 1983, 95:247-261.
21. Hall, I.R. and Kelson, A. *New Zealand J. Agri. Res.* 1981, 24:221-222.
22. Hewitt, E.J. Commomw. Bur. Hort. Plant Crops (GB) Rech. Commun., 1996, p. 22.
23. Dehne, H.W. and Backhaus, G.F. *Zt. Pfl. Krankh.* 1986, 93:415-424.
24. Vestburg, M. and Sukainen, M.U. *The Mycologist* 1992, 6:38.
25. Winsor, G.W., Hurd, R.G. and Price, D. Nutrient film technique, Vol. 5, Glasshouse Crops Research Institute, Growers Bull., 1979.
26. Mosse, B. and Thompson, J.P. *Can. J. Bot.* 1984, 62:1523-1530.
27. Zobel, R.W., Tredici, P.D. and Torrey, J.G. *Pl. Physiol.* 1976, 57:344-346.
28. Hung, L.L. and Sylvia, D.M. *Appl. Environ. Microbiol.* 1988, 54:353-357.
29. Mosse, B. *J. Gen. Microbiol.* 1962, 27:509-520.
30. Mosse, B. and Hepper, C.M. *Plant Pathol.* 1975, 5:215-223.
31. Miller-Wiideman, M.A. and Watrud, L.S. *Can. J. Bot.* 1984, 30:642-646.
32. Becard, G. and Fortin, J.A. *New Phytol.* 1988, 108:211-218.
33. Strullu, D.G. and Romand, C. *C.R. Acad. Sci.* 1987, 303:245-250.
34. Banerjee, S., Zehra, M. and Kukreja, A.K. et al. *Curr. Res. Med. Arom. Pl.* 1995, 17:348-378.
35. Chilton, M.D., Tepfer, D.A. and Petit, A. et al. *Nature* 1982, 295:432-434.
36. Tepfer, D. *Plant Microbe Interactions* 1989, 6:294-342.
37. Abdul-Khaliq and Bagyaraj, D.J. *Ind. J. Exp. Biol.* 2000, 38:1147-1151.
38. Ryder, M.H., Tate, M.E. and Kerr, A. *Pl. Physiol.* 1985, 77:215-221.
39. Morgan, A.J., Cox, P.N. and Turner, D.A. et al. *J. Plant Sci.* 1987, 49:37-49.
40. Zhipeng, Z. and Shiuchien, K. *Acta Microbiologica* 1991, 31:32-35.
41. Gerdemann, J.W. and Nicolson, T.H. *Trans. Br. Mycol. Soc.* 1963, 46:234-235.
42. Furlan, V., Barschi, H. and Fortin, J.A. *Trans. Br. Mycol. Soc.* 1980, 75:336-338.
43. Mertz, S.M., Heithaus III, J.J. and Bush, R.L. *Trans. Br. Mycol. Soc.* 1979, 72:167-169.
44. Tommerup, I.C. and Kidby, P.B. *Appl. Environ. Microbiol.* 1979, 37:831-835.
45. Elias, K. and Safir, G.R. *Appl. Environ. Microbiol.* 1987, 53:1928-1933.
46. Mugnier, J. and Mosse, B. *Phytopath.* 1987, 77:1045-1050.
47. Nuutilla, A.M., Vestberg, M. and Kauppinen, V. *Pl. Cell Rep.* 1995, 14:505-509.
48. Watrud, L.S., Heithaus III, J.J. and Jaworski, E.G. *Mycologia* 1978, 70:449-452.
49. Becard, G. and Piche, Y. *Appl. Environ. Microbiol.* 1989, 55:2320-2325.
50. White, P.R. The cultivation of animal and plant cells, 2nd ed. Ronald Press, New York, 1954, p. 258.
51. Murashige, T. and Skoog, F. *Physiol. Plant.* 1962, 15:473-490.
52. Mosse, B. and Phillips M. *J. Gen. Microbiol.* 1971, 69:157-166.
53. Baylis, G.T.S. *New. Phytol.* 1959, 58:274-280.
54. Mosse, B. *Can. J. Bot.* 1988, 66:2533-2540.
55. Chabot, S., Belrhlid, R., Chenevert, R. et al. *New Phytol.* 1992, 22:461-467.

56. St. Arnaud, M., Hamel, C., Vimard, B. et al. *Mycol. Res.* 1996, 100:328-332.
57. Diop, T.A. *These de docteur en biologie et physiologie vegetale.* Angers, France, 1995.
58. Diop, T.A., Plenchette, C. and Strullu, D.G. *Mycorrhiza* 1994, 5:17-22.
59. Declerck, S., Strullu, D.G. and Plenchette, C. *Mycol. Res.* 1996, 100:1237-1242.
60. Fortin, J.A., Bécard, G. and Declerck, S. et al. *Can. J. Bot.* 2002, 80:1-20.
61. Bi, Y., Li, X., Wang, H. et al. *Plant Soil*, 2004. 261:239-244.
62. Gianinazzi-Pearson, V., Branzanti, B. and Gianinazzi, S. *Symbiosis* 1989, 7:243-255.

5

Biomass Production, Arbuscular Mycorrhizae and Soil Plant-available P under Water Stress in Native Perennial Grasses

C.A. Busso and A.I. Bolletta[1]

Dept. Agronomía, Universidad Nacional del Sur and CERZOS (CONICET)
8000 Bahía Blanca, Argentina
[1]INTA EEA. Bordenave, PO Box 44, 8187 Bordenave, Argentina

Introduction

Precipitation determines vegetation type and species relative composition in plant communities [1]; in addition, it is a primary variable affecting soil water storage and availability. This availability is the most important factor which affects plant distribution, growth and survival in rangelands throughout the world [2]. Plant productivity can be reduced in these rangelands because of the negative effects of water stress on plant growth [2].

The generally adverse effect of high soil plant-available phosphorus levels on Arbuscular Mycorrhizae (AM) formation is well documented [3, 4], and is mainly caused by higher P concentrations in the roots [5]. In soils well supplied with phosphate, mycorrhizal plants typically grow worse than non-mycorrhizal ones, because mycorrhizae can behave as parasitic rather than mutualistic in the fungi-plant association [6]. Plant species with lower root length density are more dependent on mycorrhizal colonization for resource acquisition than species with higher root length density [7]. *Nassella clarazii* has greater root length density [8] and mycorrhizal colonization [9] than *N. tenuis* and *Amelichloa ambigua*. These species are abundant in rangelands of Central Argentina [10]. While *N. clarazii*, a late-seral, highly competitive species, is selectively consumed by domestic herbivores, *N. tenuis*, an earlier-seral and less competitive species, is an intermediate species to herbivory, and *A. ambigua* and *S. gynerioides*, both early-seral, low competitive species are only cut-off when a better forage is not available [10].

Nassella clarazii is a more competitive species than *N. tenuis* and *A. ambigua* [11-13]. Shoot growth, root proliferation, root length densities and nutrient uptake rates have been shown to be greater in *N. clarazii* than in *N. tenuis* and *A. ambigua* [11-14]. Covacevich et al. reported that AM colonization was not depressed in wheat at soil plant-available P concentrations of upto 15 ppm [15]; beyond these soil plant-available P concentrations, root colonization of wheat by AM was highly depressed. Soils in our study site had between 24 and 55 ppm soil plant-available P concentrations. The shoot and root physiological traits in *N. clarazii*, and the high soil plant-available P concentrations at the study site, suggest that high AM colonization in *N. clarazii* could act in a parasitic rather than a mutualistic way. If so, aboveground plant biomass in the late-seral, highly competitive *N. clarazii* would be expected to decrease as percentage arbuscular mycorrhizal colonization increase. On the other hand, lower values of the above mentioned physiological traits in *N. tenuis* and the early-seral, low competitive *S. gynerioides* suggest that aboveground biomass production would not diminish with increases in root percentage AM colonization.

Arbuscular mycorrhizal fungi are common mutualistic symbionts of plant roots in grasslands [16]. Arbuscular mycorrhizal associations have been found in areas that cover a wide range of soil moisture [17]. Under experimental conditions, it has been reported that arbuscular mycorrhizal colonization lessened with decreasing soil moisture availability in the perennial and annual grasses *Schizachyrium scoparium* (Michx.) Nash and *Sorghum bicolor* (L.) Moench., respectively [18, 19]. Jupp and Newman reported the effect of soil moisture on soil plant-available phosphorus availability [20]. Plant phosphorus uptake and transport, and diffusion coefficients of ^{32}P, were significantly reduced as soil moisture content (SMC) decreased [21-25]. Root phosphorus is also added to the fresh organic phosphorus pool upon its death and/or incorporation into the soil. Decomposition of fresh (and stable) organic matter may result in net mineralization of organic phosphorus [26]. In addition, plant growth rate reductions under water stress [27] directly influence phosphorus uptake. As a result, phosphorus concentrations in the soil solution have increased when SMC decreased [28].

Various morphological and physiological characteristics have been associated with an effective nutrient acquisition and plant competition. Plants of low productive environments characterize by low nutrient acquisition rates, nutrient retention and low growth rates to maintain a balance between availability and demand [29]. In these less productive environments, a high soil resource competition is assumed. As a result, high infection frequencies by AM [30] are common root characteristics of semiarid rangeland species. However, linear relationships between soil versus root characteristics and plant growth should not necessarily be expected [31, 32], and these relationships are often difficult to identify in natural systems [33]. In addition, roots of various species may show a high morphological and physiological plasticity which allow them an increased exploration of heterogeneously-distributed soil resources [32].

The following general hypotheses will be tested in this field study: Total annual production of biomass is reduced under severe water stress (imposed during the vegetative plus internode elongation developmental stages) in all study species. However, total annual aboveground plant biomass under all study water levels, and that produced during the initial regrowth period in the following year, when all plants are released from water stress, are greater in *N. clarazii* than in *N. tenuis*. Aboveground biomass production of *S. gynerioides* is highly stimulated by greater than lower SMC. High mycorrhizae percentages in soils with high soil plant-available P concentrations decrease aboveground biomass production in the late-seral, high competitive species *N. clarazii*, but not in the earlier-seral, comparatively less competitive species *N. tenuis* and *S. gynerioides*. Root AM colonization decrease with decreasing SMC in all study species. Finally, soil plant-available P concentrations under the canopy of all study species increase meanwhile SMC decreases.

Objectives of this work included to determine (1) the effects of different SMC (water stress, rainfed or irrigated conditions) on aboveground biomass production, colonization levels by AM on *N. clarazii* and *N. tenuis* in competition with plants of *S. gynerioides*, and available soil plant-available P concentrations, and (2) the relationships among AM colonization levels, plant biomass production and soil plant-available phosphorus concentrations. As emphasized [34], these studies will contribute to increase our understanding of the interdependency of the responses between shoot and root tissues in the study species.

Materials and Methods

Study Site

This research was conducted during 1995, 1996 and early 1997 in proximities of the Departamento de Agronomía, Centro de Recursos Renovables de la Zona Semiárida (CERZOS) in Bahía Blanca city (38° 48′ S, 62° 13′ W). Plots were established in a typical Haplustol soil (L. Sánchez, Dept. Agronomía, UNS, personal communication). In this soil, five horizons are distinguished (A-AC-C-2Ck-3Ckm); a calcium carbonate horizon ($CaCO_3 > 300$ gr kg^{-1}) is present at 1.80 m depth, and pH is 6.6-7 in the root zone [35].

Rainfall, temperature and potential evapotranspiration data during the study period were obtained in a meteorological station located 100 m apart from the experimental plots (Figure 1). During 1995, 1996 and early (January-March) 1997 rainfall was 447.2, 621.3 and 285 mm, respectively. Rainfall distribution

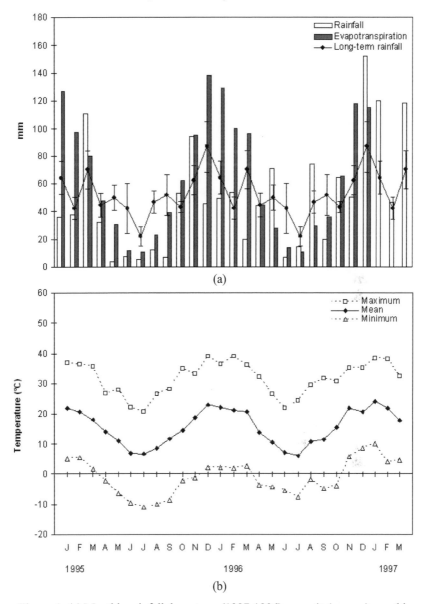

Figure 1: (a) Monthly rainfall, long-term (1987-1996) mean (± 1 s.e.m.) monthly rainfall, and mean monthly potential evapotranspiration and (b) Absolute minimum and maximum and mean monthly air temperatures to 0.25 m above the soil surface during 1995, 1996 and early 1997. Measurements were taken using a meteorological station located 100 m away from the experimental plots.

was seasonal in 1995, with peaks in fall and spring. Rainfall distribution was more uniform during 1996. Absolute minimum and maximum, and mean monthly temperatures were similar in both years. Mean minimum temperature was 7°C during June and July, and mean maximum temperature was 22 to

24°C in January. Potential evapotranspiration was also similar in both years. In general, 1996 was a wetter year than 1995.

Plant Material

Research was conducted on the late-seral, highly competitive *Nassella clarazii*, the comparatively earlier-seral, comparatively less competitive *Nassella tenuis* and the early-seral, poorly competitive *Stipa gynerioides* [36]. These are three C_3 perennial grass species native to the Distrito Fitogeográfico del Caldén [37]. Their growing cycle occurs during fall, winter and spring. The first two species have a high forage value (desirable, preferred, palatable) [37] while *S. gynerioides* is a non-preferred species. This species was included in the experimental plots because it is the most abundant, undesirable perennial grass at the south of such District [38].

Experimental Design

Between December 1993 and April 1994, 28 experimental plots (1.8 × 1.8 m) were established in the field on unplowed, weeded soil. Plants were obtained from a 20 year-exclosure to domestic animals located southeast of La Pampa Province (38° 45' S, 63° 45' W). Within each plot, transplants were placed 30 cm apart from one another in seven horizontal and vertical rows such that each plant of *N. clarazii* or *N. tenuis* was surrounded by four plants of *S. gynerioides*. Disposition of plants within a uniform matrix contributes to reduce potentially confounding effects on plant responses as a result of plant competition. A total of 1372 transplants were used for the whole study. Crown-level plant diameters (n=56) were similar among species at time of transplanting: 13.47±0.56 cm (mean±1 SE) for *N. clarazii*, 10.02±0.51 cm for *N. tenuis*, and 12.27±0.61 cm for *S. gynerioides*. All tussocks of *N. clarazii* and *N. tenuis* were hand-clipped to a 5-cm stubble height in January 1995, during the plant quiescent period. From a total of 28 experimental plots, eight were randomly assigned to the irrigated and eight to rainfed treatments, and four plots to each of the water stress treatments (vegetative, internode elongation, and vegetative plus internode elongation).

Water Levels

Plants were exposed to rainfed, irrigated or water stress conditions. Rainfed plots received rainfall all year round (Figure 2). A drip irrigation system watered the irrigated plots, which were additionally rainfed. Soil tensiometers installed in the irrigated plots allowed watering of these plots to saturation whenever they reach 60% of field capacity. Periods of irrigation and imposition of water stress during 1995 and 1996 are depicted in Figure 2. Transparent plastic sheets covered the water-stressed plots whenever rain fell at periods when these species are often exposed to water stress in their native environment [10]: vegetative

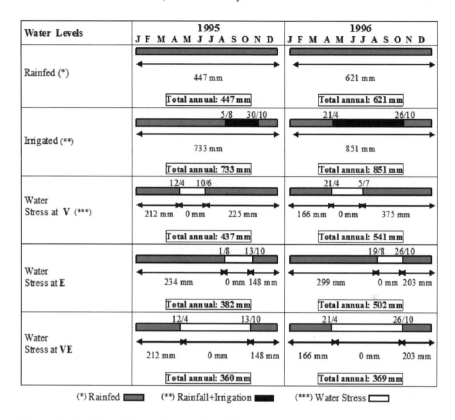

Figure 2: Periods of imposition of the different water inputs at the vegetative (V), internode elongation (E) or both (VE) phenological stages in 1995 and 1996. Numbers below horizontal, bold lines are rainfall fallen in the rainfed, and water stress treatments, or rainfall + irrigation in the irrigated treatment. Total annual precipitation is indicated for each year and water level within rectangles. Numbers immediately above horizontal bars represent the beginning and end, respectively, of imposition of any given water level. Black, grey or white horizontal bars represent irrigated, rainfed or water stress conditions, respectively.

or early internode elongation or both phenological periods (Figure 2). Water-stressed plots were surrounded with plastic sheets up to 1.8 m soil depth to prevent lateral movement of water into these plots. All 28 experimental plots received 313.7 mm from mid-October 1995 to late-April 1996, and 487.8 mm from late-October 1996 to March 1997. Water-stressed plots were thus alleviated from water stress during these periods by receiving natural rainfall.

Sampling Procedures

Leaf Water Potentials

Leaf water potential was determined periodically at mid-day in all treatments to provide a measure of plant water status during the study period. Measurements were done using a pressure-chamber on sunny days only between 1200 and 1300 h. Youngest, fully expanded leaf blades were taken for these measurements using one tiller per species within each replicate plot and sampling date. From excision to end of each determination, leaves were cut one at a time and maintained in a plastic bag to reduce water loss [39].

Aboveground Biomass Production

In January (summer) 1996, the dormant season, all plants of the desirable species (*N. clarazii* and *N. tenuis*) were defoliated to 5-7 cm stubble under all water levels to measure new plant growth from that height. On 21 December of the same year, all plants were defoliated to that stubble height within each plot to quantify biomass production during 1996. With this purpose, 4 to 8 plants were sampled within each water level. Defoliation on 21 December 1996 also allowed to measure biomass production from this date to 28 February 1997. At this date, biomass production of all three species was evaluated, defoliating for the first time during the study of the plants of *S. gynerioides*. *Stipa gynerioides* remained undefoliated during the study because this species is only cut off when there is not better forage [37]. Defoliated material was oven-dried to 60° C during 72 hrs and weighed.

Soil Available Phosphorus and AM

No P supplements were provided during the study. Soil plant-available P concentrations at the outset of the experiment (1995) were not determined because parallel, labour-intensive studies were conducted on these plots at that time. We recognize that this limits use of plant and soil plant-available P data as a measure of plant productivity and the influence of root characteristics including AM. However, during the study period (1996-early 1997), samplings for soil available phosphorus determinations were not only conducted under water stress and irrigated but also rainfed (untreated control) conditions.

A total of 460 soil plus root samples were obtained between 0-15 cm soil depth using a soil corer (8.4 cm diameter, 15 cm height: 831.3 cm^3 volume) during April, June, September and October in 1996, and February in 1997. Samples were obtained diagonally from the plant periphery to the plant centre to assure that sampled soil plus roots corresponded to the sampled plant. One plant of each species was used per replicate at each sampling date. Roots were obtained after washing soil samples through a 60 mesh screen [40], and they were maintained at 4°C in a solution of formaldehyde, glacial acetic acid and ethanol [41] for AM determinations. A parallel, similar soil sampling was conducted on each plant for available phosphorus determinations.

Soil available phosphorus was determined by the method of Bray and Krutz [42]. Soil was air-dried and then screened through a mesh of 0.5 mm, and 2.5 g soil was weighed. This sample was placed within a tube which contained 20 ml of Bray and Kurtz's extractant. The tube was agitated during 5 min. at 190 agitations per minute. Content of the tube was filtered, and P (ppm) was determined by colorimetry using a UV-Visible Recording Spectrophotometer, UV-2100 Shimadzu. Washed roots were cut into 15 mm segments, cleared and stained for determination of mycorrhizae colonization at 100-400X magnification [43]. Three fields on each of thirty root segments were scored for presence or absence of hyphae, vesicles and arbuscules for each plant.

Statistical Analysis

Leaf Water Potentials

Leaf water potentials were analysed using a three-way ANOVA (five water levels × three species × four sampling dates: April, June and September 1996, and February 1997) in split plot. Soil water levels were the main factor applied to randomly distributed plots, in an unbalanced but proportional manner; there were four replicates for the irrigated and rainfed treatments, and two replicates for each of the water stress treatments (vegetative, internode elongation and vegetative plus internode elongation). There were plants of the three species within each plot; one plant of each species was assigned to be sampled within each sampling date. Secondary factors were sampling dates and species. Interactions were open to evaluate the effects of water levels, dates and species. Since interactions involved water levels, species and sampling dates, comparisons were conducted (1) among water levels for each sampling date and (2) among sampling dates for each water level on each of the study species individually. Means were compared by LSD at 5%, when the F tests indicated that the variables were different at 5% [44].

Aboveground Biomass Production

Biomass production in 1996 and that produced between 21 December 1996 and 28 February 1997 were individually analyzed using a two-way ANOVA (five water levels × three species) in split plot. Soil water levels were the main factor, applied to randomly distributed plots, in an unbalanced but proportional manner. There were plants of the three species within each plot, assigning one plant of each species for analysis within each sampling date. Secondary factors were the species.

Available Soil Plant-available P

At first, soil plant-available phosphorus concentration data were analyzed using a three-way split plot ANOVA (five water levels × three species × five sampling dates: April, June, September and October 1996, and February 1997). Soil water levels acted as main factors, applied to randomly distributed plots, in a proportional but unbalanced manner. Eight replicates were used for the irrigation and rainfed treatments, and four replicates were utilized for each of the water stress treatments (vegetative, internode elongation, vegetative plus internode elongation). Plants of the three species were within each plot, assigning one plant of each species for analysis at each sampling date. Secondary factors were sampling dates and species. Within each plot, plants were assigned for sampling previous to the sampling dates. This allowed avoiding measurements of available soil plant-available phosphorus corresponding to nearby plants previously sampled. However, this rigid scheme did not allow replacement of lost plants (i.e., plants which died as a result of treatment application) during the study. Because of this, it was necessary to adapt the statistical analysis when the loss of sampling units resulted in an unbalance not proportional among the species within each plot. This phenomenon was mainly presented in June and September.

In this way, we have a 'Design 1' using a three-way ANOVA for those months where information was complete: April, October and February. Months with missing data (June and September) were analyzed with a 'Design 2': a split-plot two-way ANOVA with the same main factor (soil water levels) and a unique secondary factor (the species). Uncompleted plots, which lack information on the three species, were eliminated to apply this analysis, leaving an unbalanced, proportional design. Interactions were open to evaluate the effects of water levels, dates and species. Means were compared with Fisher's protected LSD at 5% when the F test indicated that the variables differed at that significance level [44].

Relationships among Study Variables

According to the available variables on each sampling date and for each species [% AM: Percentage AM colonization; Soil plant-available P: available soil plant-available phosphorus; Biomass: Annual aboveground biomass (1996) and that produced between 21 December 1996 and 28 February 1997; Water pot: Leaf water potential] correlation coefficients (*r*) were calculated according to the scheme shown in Figure 3. Correlation coefficients obtained at the

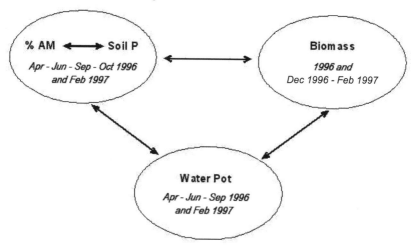

Figure 3: Scheme showing the variables used for calculation of the correlation coefficients among them for each species at the shown sampling dates.

different dates were compared using statistics with Chi-square distribution based on the 'Z' transformation of Fisher, where r is the correlation coefficient, k is the number of correlation coefficients being compared, and n is the number of data pairs which are used to calculate each correlation coefficient. Used formulae were as follows:

$$X^2 = \sum_{i-1}^{k}(n_i - 3).Z_i^2 - \frac{\left[\sum(n_i - 3).Z_i\right]^2}{\sum(n_i - 3)} \qquad Z = \frac{1}{2}\ln\left(\frac{1+r}{1-r}\right)$$

If correlation coefficients did not differ, a correlation coefficient combining all dates was calculated. If differences were detected, pair comparisons were effected using LSD at $p<0.05$. According to the results, dates were grouped to find the corresponding combined correlation coefficients. Its significance (Q=0 versus Q≠0) was tested on each of these using a normal distribution test. If any correlation coefficient remained isolated (i.e., was not the result of combining several dates), its significance was tested with a t test [45]. Only statistically significant, biologically meaningful correlations will be reported ($p \leq 0.05$).

Results

Leaf Water Potentials

In April 1996 and February 1997, leaf water potentials of *N. clarazii*, *S. gynerioides* and *N. tenuis* did not differ ($p>0.05$) among soil water levels (Figure 4). However, leaf water potentials were lower ($p<0.05$) under water stress at the (1) vegetative and (2) vegetative plus internode elongation developmental stages than under rainfall or irrigated conditions in June 1996 for all three species (Figure 4). Also, leaf water potentials of *N. clarazii*, *N. tenuis* and *S. gynerioides* were lower ($p<0.05$) under water stress in the internode elongation and vegetative plus internode elongation stages than under the other water levels in September of the same year. Lowest leaf water potentials ($p<0.05$) were found in September 1996 and February 1997 under rainfed and irrigated conditions in *N. clarazii* and *S. gynerioides* (data not shown). In June 1996, leaf water potentials were lower ($p<0.05$) than in September of the same year when water stress occurred at the vegetative stage in *N. clarazii* and *S. gynerioides*. In September 1996, leaf water potentials were lower ($p<0.05$) than values in June in the internode elongation stage, and greater ($p<0.05$) than values in June at the vegetative stage in all three species (Figure 4).

Aboveground Biomass Production

Plants of *N. clarazii* reached a greater ($p<0.05$) annual aboveground biomass under irrigated, rainfed and water stress conditions at the vegetative or internode elongation developmental stage than under water stress in both phenological

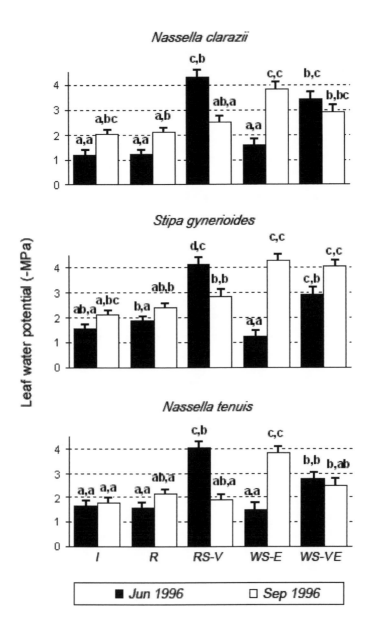

Figure 4: Mid-day leaf water potential (-MPa) on plants of *N. clarazii*, *N. tenuis* and *S. gynerioides* which were exposed to irrigated (I), rainfed (R) or water stress (WS) conditions at the vegetative (V), internode elongation (E) or both (VE) phenological stages during June and September 1996. Values in the Y axis are represented as absolute values. Each histogram is an average of *n*=2-4. Vertical bars represent one s.e.m. Different letters to the left of the comma indicate significant differences (*p*<0.05) among water levels, and those to the right of the comma indicate significant differences (*p*<0.05) among sampling dates.

stages in 1996 (Figure 5). Plants of *N. tenuis*, however, have a similar ($p>0.05$) biomass under all water levels. *Nassella clarazii* showed a greater ($p<0.05$) annual biomass than *N. tenuis* under irrigated and rainfed conditions, and under water stress in the vegetative or internode elongation stage in 1996 (Figure 5). In February 1997, *N. clarazii* showed a greater ($p<0.05$) biomass under water stress in the vegetative stage than under the remaining water levels (Figure 5). In *Nassella tenuis*, there were no differences ($p>0.05$) among water levels (Figure 5). *Nassella clarazii* reached a greater ($p<0.05$) biomass than *N. tenuis* under water stress conditions in the vegetative stage. *Stipa gynerioides* had a similar ($p>0.05$) biomass production under all water levels in 1997 (Figure 6).

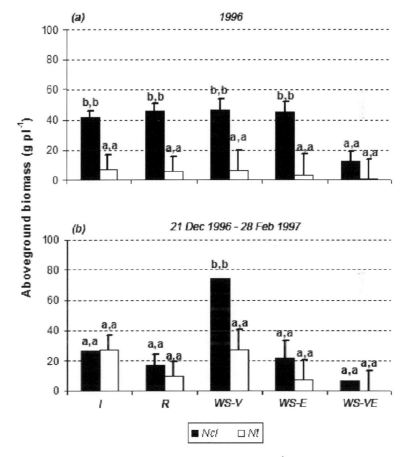

Figure 5: Annual aboveground biomass [(A), g plant^{-1}] and aboveground biomass produced between 21 December 1996 and 27 February 1997 [(B), g plant^{-1}] on plants of *N. clarazii* (Ncl) and *N. tenuis* (Nt) exposed to irrigated (I), rainfed (R), or water stress (WS) conditions at the vegetative (V), internode elongation (E) or both phenological stages (VE). Each histogram is the mean of n=4-8. Vertical bars represent one s.e.m. Different letters to the left of the comma indicate significant differences ($p<0.05$) among water levels, and those to the right of the comma indicate significant differences ($p<0.05$) between species.

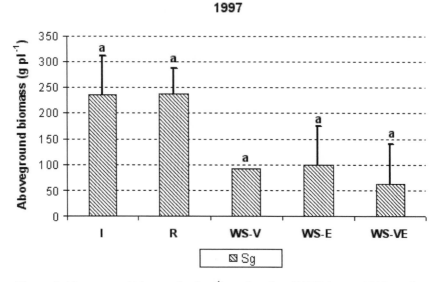

Figure 6: Aboveground biomass (g plant[-1]) produced until 27 February 1997 on plants of *S. gynerioides* (Sg) exposed to irrigated (I), rainfed (R), or water stress (WS) conditions at the vegetative (V), internode elongation (E) or both phenological stages (VE). Each histogram is the mean of *n*= 4-5. Vertical bars represent one s.e.m. Equal letters above histograms indicate lack of significant differences (*p*>0.05) among soil water regimes.

Available Soil Plant-Available P

When soil plant-available phosphorus concentration was analyzed on average for the April, October and February sampling dates, there was an interaction (*p*<0.05) between water levels, dates and species. This implies that the response in the various water levels was different (*p*<0.05) during these months in all three species. Because of this, each species was analyzed individually comparing dates and water levels (Table 1). In *N. clarazii* and *N. tenuis*, there was a different response in the different water levels in April, October and February (*p*<0.05) (Table 1). In April 1996, soil plant-available phosphorus concentration was greater (*p*<0.05) than in October of the same year and February 1997 where *N. clarazii* had been exposed to water stress during the vegetative developmental stage. On the other hand, soil plant-available phosphorus concentrations were greater (*p*<0.05) in October 1996 than in April 1996 and February 1997 when *N. clarazii* and *N. tenuis* were exposed to water stress in both developmental stages. In all sampling dates, each species had a similar (*p*>0.05) soil plant-available phosphorus concentration under all water levels (Table 1). Soil plant-available phosphorus concentrations under *S. gynerioides* were similar (*p*>0.05) among water levels and sampling dates (Table 1).

In June 1996, soil plant-available phosphorus concentrations were similar among species (*p*>0.15) and water levels (*p*>0.80) (Table 2). However, in

Table 1. Soil phosphorus concentration (ppm) under the canopy of *N. clarazii* (*Ncl*), *S. gynerioides* (*Sg*) and *N. tenuis* (*Nt*) plants exposed to irrigated (I), rainfed (R) or water stress (WS) conditions at the vegetative (V), internode elongation (E) or both phenological stages (VE) during April and October 1996, and February 1997

	April 1996			October 1996			February 1997		
	Ncl	*Sg*	*Nt*	*Ncl*	*Sg*	*Nt*	*Ncl*	*Sg*	*Nt*
I	$44.2^{a,ab}$	$36.8^{a,a}$	$30.8^{a,a}$	$43.9^{a,a}$	$43.7^{a,a}$	$33.2^{a,a}$	$42.6^{a,a}$	$38.8^{a,a}$	$39.5^{a,a}$
R	$40.3^{a,a}$	$36.9^{a,a}$	$30.8^{a,a}$	$48.2^{a,ab}$	$43.3^{a,a}$	$32.7^{a,a}$	$40.8^{a,a}$	$39.0^{a,a}$	$40.1^{a,a}$
WS-V	$49.6^{a,b}$	$34.3^{a,a}$	$34.5^{a,a}$	$40.9^{a,a}$	$36.3^{a,a}$	$29.0^{a,a}$	$41.4^{a,a}$	$28.2^{a,a}$	$34.0^{a,a}$
WS-E	$43.1^{a,ab}$	$26.5^{a,a}$	$34.7^{a,a}$	$45.9^{a,a}$	$32.0^{a,a}$	$32.5^{a,a}$	$38.9^{a,a}$	$31.7^{a,a}$	$31.8^{a,a}$
WS-VE	$37.7^{a,a}$	$47.3^{a,a}$	$33.7^{a,a}$	$55.0^{a,b}$	$36.3^{a,a}$	$46.8^{a,b}$	$44.5^{a,a}$	$35.1^{a,a}$	$34.8^{a,a}$

Each value is the mean of $n=2-8$; Different letters to the left of the comma indicate significant differences ($p<0.05$) among soil water levels; Different letters to the right of the comma indicate significant differences ($p<0.05$) among sampling dates.

Table 2. Soil phosphorus concentration (ppm) under the canopy of *N. clarazii* (*Ncl*), *S. gynerioides* (*Sg*) and *N. tenuis* (*Nt*) plants exposed to irrigated (I), rainfed (R) or water stress (WS) conditions at the vegetative (V), internode elongation (E) or both phenological stages (VE) during June and September 1996

	June 1996			September 1996		
	Ncl	*Sg*	*Nt*	*Ncl*	*Sg*	*Nt*
I	$38.4^{a,a}$	$42.0^{a,a}$	$34.7^{a,a}$	$37.8^{a,a}$	$35.0^{b,a}$	$35.1^{a,a}$
R	$40.2^{a,a}$	$39.6^{a,a}$	$35.7^{a,a}$	$41.2^{ab,b}$	$38.2^{b,b}$	$32.1^{a,a}$
WS-V	$42.1^{a,a}$	$40.4^{a,a}$	$35.1^{a,a}$	$43.6^{ab,b}$	$23.9^{a,a}$	$30.1^{a,a}$
WS-E	ND	ND	ND	$46.8^{b,b}$	$30.7^{ab,a}$	$31.7^{a,a}$
WS-VE	$48.7^{a,a}$	$50.3^{a,a}$	$36.1^{a,a}$	$47.7^{ab,a}$	$44.6^{b,a}$	$34.9^{a,a}$

Each value is the mean of $n=2-8$; Different letters to the left of the comma indicate significant differences ($p<0.05$) among soil water levels; Different letters to the right of the comma indicate significant differences ($p<0.05$) among species; ND: Not Determined.

September, soil below *N. clarazii* plants had a greater ($p<0.05$) phosphorus concentration when plants were exposed to water stress under the internode elongation developmental stage than under irrigation (Table 2). Under *S. gynerioides*, soil plant-available phosphorus concentrations were lower ($p<0.05$) under water stress in the vegetative stage than under irrigated, rainfed and water stress conditions in the vegetative plus internode elongation developmental stages (Table 2). There were no differences ($p>0.05$) in soil plant-available phosphorus concentrations among water levels in *N. tenuis*. Soil under the canopies of *N. clarazii* and *S. gynerioides* had a greater ($p<0.05$) phosphorus concentration than that under the canopy of *N. tenuis* under rainfed conditions (Table 2). However, soil plant-available phosphorus concentrations were greater ($p<0.05$) under plants of *N. clarazii* than under those of the other two species under water stress conditions in the vegetative or internode elongation developmental stage (Table 2).

Relationships between Root and Shoot Variables

Total annual biomass production in 1996 and the subsequent R during 21 December 1996-28 February 1997 after water stress was released, were negatively correlated ($p<0.05$) with percentage AM colonization (Figure 7) in *N. clarazii* under all water levels. These relationships were not significant ($p>0.05$) for either *N. tenuis* or *S. gynerioides* under all water regimes. In *N. tenuis*, leaf water potential (X, -MPa) and AM colonization (Y, %) showed a negative correlation ($Y=34.217 - 6.3323X$, $r=0.699$, $p=0.05$, $n=8$) in June 1996. Percentage AM colonization (Y, %) of *N. clarazii* correlated negatively ($Y=181 - 6.62X + 0.0677X^2$, $r=0.51$, $p=0.003$, $n=42$) with soil plant-available phosphorus availability (X, ppm) in April and June 1996.

Discussion

The scope of data interpretation might be limited in this study because samplings were conducted for only one growing season (January 1996-February 1997). However, 1996 appeared to be a rather typical, average year at the study site. For example, rainfall during this year (621 mm) was similar to the long-term mean (633.9 ± 58.3 mm, mean$\pm$1e.e.) at the same research area during 1987-1996. No long-term evapotranspiration data were available. In general, we were successful in imposing water stress conditions at the field in all three species in comparison to irrigated plots. Lack of significant differences ($p>0.05$) in leaf water potentials in April 1996 and February 1997 was due to the fact that (1) water stress was imposed at the end of April 1996, and (2) all plots were exposed to natural rainfall after 26 October 1996, respectively. Leaf water potentials found in June and September are similar to results reported in these and other perennial grasses [46, 47].

In arid and semiarid environments, soil water availability affects plant growth and survival in the plant community [2]. Water stress, or years with precipitations below the annual mean, decrease dry matter production in several perennial grass species [48]. Plants of *N. clarazii* and *N. tenuis* were not shaded by plants of *S. gynerioides* under water stress, even though these plants remained undefoliated until the end of the study. In agreement with the general hypothesis, biomass production of *N. clarazii* was lower under water stress in the vegetative plus internode elongation stage than under the remaining soil moisture regimes. Similar results were obtained [2, 48] in other perennial grass species. However, despite using a strict experimental plot setup to obtain plant water stress under field conditions, annual biomass in *N. tenuis,* and that accumulated by *S.*

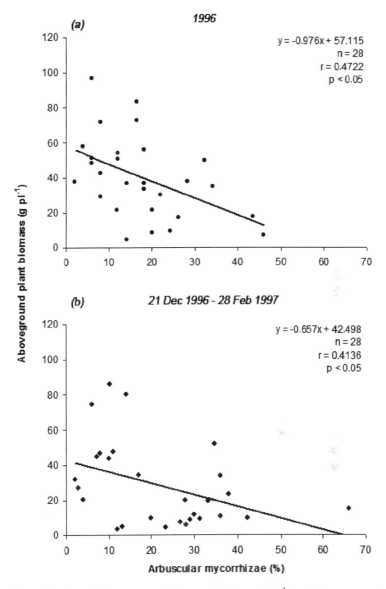

Figure 7: Annual aboveground biomass [(A), g plant⁻¹] and aboveground biomass produced between 21 December 1996 and 28 February 1997 [(B) g plant⁻¹] versus arbuscular mycorrhizae colonization percentage in *N. clarazii*. Each symbol represents a single observation from the pool of data obtained under irrigated, rainfed or water stress conditions.

gynerioides, were similar among soil water levels, which disagrees with the general hypothesis. This calls the attention of the difficulty of imposing water stress conditions at the field in an environment where annual precipitation may be higher than 600 mm. Similar results were obtained by Asay and Johnson

[49-51]. *Nassella clarazii* showed a higher average degree of osmotic adjustment (−2.72 MPa) than *N. tenuis* (−1.94 MPa) and *S. gynerioides* (−2.34 MPa) under water stress in the internode elongation stage of development [36]. This drought tolerance mechanism will very likely help these species to keep growing in the face of declining leaf water potentials [39], and would help explain, at least partially, why aboveground biomass production of these species, except that on *N. clarazii*, was similar under all soil water regimes. The lower biomass production of *N. clarazii* under water stress in the vegetative plus internode elongation than in the other developmental stages suggests that this species may be more sensitive than *N. tenuis* and *S. gynerioides* to long-term water stress.

In 1996, annual biomass production was greater in *N. clarazii* than in *N. tenuis* under irrigated and rainfed conditions and under water stress in the vegetative or internode elongation developmental stage. These results are consistent with those obtained under rainfed conditions [14, 52, 53]. This greater biomass production in *N. clarazii* than in *N. tenuis* may be due, at least partially, to the well known greater relative growth rates and competitive ability in the first than in the second species [11-14, 52, 53].

Increases in percentage AM colonization under conditions of high soil plant-available P concentrations resulted in aboveground biomass decreases in the late-seral, highly competitive perennial grass *N. clarazii*. This species has high root length densities [13, 54, 55]. These high root length densities appear to make this species less dependent on mycorrhizal colonization for resource acquisition. Species with lower root length densities were reported to be more dependent on AM colonization for resource acquisition, such as P, than species with higher root length densities [7]. The relationships between AM colonization percentage and aboveground biomass production were not significant in the earlier-seral, less competitive *N. tenuis* and *S. gynerioides*. These species have a much lower root length density than *N. clarazii* [13]. Covacevich et al. [15] reported that percentage arbuscular mycorrhizal colonization in wheat was highly depressed in soils with greater than 15 ppm plant-available P. In our study, where soils had between 10 and 60 ppm plant-available P, the relationship between aboveground biomass production and percentage arbuscular mycorrhizal colonization appeared to be parasitic rather than mutualistic in the late-seral, highly competitive perennial grass *N. clarazii*. However, no such relationship was found for the earlier-seral, less competitive perennial grass species *N. tenuis* and *S. gynerioides*.

Water stress can reduce AM colonization in several grass species, which is in part associated to stress intensity [17, 19]. AM colonization significantly decreased in *N. tenuis* when leaf water potentials were reduced in June, a fact that we could not show for *N. clarazii* and *S. gynerioides*. Several studies on the relationships between soil and plant variables have found significant differences for these relationships in only one or some, but not all, study species [13, 56]. They showed that relationships between any of their study variables were species-specific.

A negative correlation was found between percentage mycorrhizae colonization and available soil plant-available phosphorus concentrations in plants of *N. clarazii* during the first sampling dates. We were unable to find this relationship in *N. tenuis* and *S. gynerioides*. The reason because mycorrhizal infection should be greater under phosphorus deficient conditions is not clear. Jasper et al. [57] have suggested that high carbohydrate level in the roots is an important prerequisite for good mycorrhizal infection. This is consistent with the observation that plants poorly supplied with phosphorus have higher root carbohydrate levels than plants adequately supplied with phosphorus [58]. Jackson and Caldwell had already emphasized that linear relationships between soil versus root characteristics should not necessarily be expected [31, 32]. In addition, it has been reported that these relationships are often difficult to identify in natural systems [33]. The result obtained in the late-seral perennial grass suggests a strategy that would allow this species avoid changes from mutualism to parasitism in the plant-mycorrhizae fungi relationship. Similar results were found in plants of *N. clarazii* [13].

It has been demonstrated that soil plant-available phosphorus has a direct effect on AM colonization [59]. Correlation coefficients obtained in *N. clarazii* between the dependent and independent variables are greater than those reported [56] for the relation between available soil plant-available P and colonized root length in *Schizachyirium scoparium*. Cool season grasses, like those in our study, generally have well-developed, highly branched root systems and are only weakly dependent on AM fungi for nutrient uptake [60, 61]. In addition to SMC, there are other factors which influence the plant-fungi symbiotic relationship, such as the physiological characteristics of the grasses [17, 62]. This plasticity in the response of AM could be a relevant cause of its persistence and importance in native ecosystems [17]. Plants poorly colonized by AM could receive nutrients through hyphae connections with plants of the same or different species. This type of hyphae connections, for example, contributes to find variations in AM colonization levels through time, and lack of correlation between different variables under field conditions. Allen et al. found a great difference in the AM activity among years, independently of various water levels and defoliation treatments [17].

Nassella clarazii showed similar or greater, but not lower, soil plant-available phosphorus concentrations under water stress than in the other soil water regimes. These findings partially agree with the general hypotheses. Also, soil plant-available P concentrations under *S. gynerioides* plants were almost double when plants were exposed to long-term (vegetative + internode elongation) than short-term (vegetative) water stress. Phosphorus mineralization from organic matter depends on soil moisture and temperature [63]. Lower self-diffusion coefficients of ^{32}P have been reported as soil water content decreased [22]. Many authors attributed increases in available soil plant-available P concentrations under water stress to a lower phosphorus uptake as soil water contents decreased [23, 24, 64, 65]. In addition, microbe and root phosphorus can be added to the fresh organic phosphorus pool upon their death and/or incorporation into the soil; subsequent decomposition of fresh organic matter may result in net mineralization of organic phosphorus [26]. Thus, we can envision at least two mechanisms by which soil plant-available phosphorus concentrations may increase with decreasing soil moisture. First, phosphorus is likely to become more concentrated in solution simply because soil moisture decreases. Second, the entire labile fraction of soil plant-available P may appear to increase relative to moist soils because drying soils result in lower rates of phosphorus diffusion and uptake; in moist soils, uptake depletes labile phosphorus. Finally, microbes and roots may liberate more plant available-phosphorus. Our results are similar to those reported [66] in a wheat crop under water stress. A continued root turnover in *N. clarazii* and *N. tenuis* under water stress conditions might contribute to maintain high levels of soil available phosphorus [67].

Available soil plant-available phosphorus concentrations were greater under the canopy of the late seral, palatable perennial grass *N. clarazii* than under the canopy of the comparatively earlier seral, palatable *N. tenuis* and the early seral, unpalatable *S. gynerioides* under the reported rainfed and/or water stress conditions. A similar response was found [68] under rainfed conditions. In their study, soil available phosphorus was greater at sites dominated by *P. ligularis*, a late-seral, palatable perennial grass, than at sites dominated by *S. tenuissima*, an unpalatable tussock grass. This would be the result, at least in part, of differences in litter and root chemical composition between species; litter and root decomposition was faster in *P. ligularis* (low C:N ratio) than in *S. gynerioides* and *S. tenuissima* (high C:N ratio) [69,70].

This study demonstrated that morphological (i.e., biomass production) and physiological (leaf water potential, AM colonization) study plant responses were very plastic, and appear to respond to the interaction of biotic and environmental characteristics that we do not yet fully understand. It also demonstrated that biomass production of late-seral, highly competitive perennial grasses, but not that of earlier-seral, less competitive perennial grass species, may not benefit by increasing percentage AM colonization under high soil plant-available P concentrations. The high root length densities of the late-seral perennial grasses may be one of the reasons to explain that increased fungi-host interactions resulted in aboveground biomass decreases in soils with high plant-available P, making parasitic rather than mutualistic the fungi-host association.

References

1. Fernández, O.A. and Busso, C.A. *In:* Case Studies of Rangeland Desertification. Arnalds, O. and Archer, S. (eds), Agric Res Inst, Report No. 200, Reykjavik, 1999, pp. 41-60.

2. Brown, R.W. *In:* Wildland Plants: Physiological Ecology and Developmental Morphology. Bedunah, D.J. and Sosebee, R.E. (eds), Soc. Range Manage., Denver, 1995, pp. 291-413.
3. Abbott, L.K., Robson, A.D. and De Boer, G. *New Phytol* 1984, 97:437-446.
4. Bååth, E. and Spokes, J. *Can J Bot* 1989, 67:3227-3232.
5. Sanders, F.E. *In:* Endomycorrhizas. Sanders, F.E., Mosse, B. and Tinker, P.B. (eds), Academic Press Ltd., London, 1975, pp. 261-276.
6. Fitter, A.H. and Hay, R.K.M. Environmental physiology of plants, Academic Press Inc., New York, 1983.
7. Koide, R.T. and Li, M. *Oecologia* 1991, 85:403-412.
8. Busso, C.A. and Bolletta, A.I. *Interciencia* 2007, 32:206-212.
9. Busso, C.A., Bolletta, A.I., Flemmer, A.C. et al. *Ann Bot Fen* 2008, 45:435-447.
10. Busso, C.A. *J Arid Environ* 1997, 36:197-210.
11. Saint Pierre, C., Busso, C.A., Montenegro, O.A. et al. *Plant Ecol* 2002, 165:161-169.
12. Saint Pierre, C., Busso, C.A., Montenegro, O.A. et al. *J Range Manage* 2004, 57:82-88.
13. Saint Pierre, C., Busso, C.A., Montenegro, O.A. et al. *Interciencia* 2004, 29:303-310.
14. Saint Pierre, C., Busso, C.A., Montenegro, O.A. et al. *Can J Plant Sci* 2003, 84:195-204.
15. Covacevich, F., Sainz Rozas, H.R., Barbieri, P. et al. *Ccia Suelo* 2005, 23:39-45.
16. Allen, M.F. The ecology of mycorrhizae. Cambridge University Press, Cambridge, 1991.
17. Allen, M.F., Richards, J.H. and Busso, C.A. *Biol Fert Soils* 1989, 8:285-289.
18. Cerligione, L.J., Liberta, A.E. and Anderson, R.C. *Can J Bot* 1988, 66:757-761.
19. Auge, R.M., Stodola, A.J.W., Ebel, R.C. et al. *J Exp Bot* 1995, 46:297-307.
20. Jupp, A.P. and Newman, E.I. *J Appl Ecol* 1987, 24:979-990.
21. Fawcett, R.G. and Quirk, J.P. *Austr J Agric Res* 1962, 13:193-205.
22. Mahtab, S.K., Godfrey, C.L., Swoboda, A.R. et al. *Soil Sci Soc Am J* 1971, 35:393-397.
23. Sánchez, P.A. and Briones, A.M. *Agron J* 1973, 65:226-228.
24. Marais, J.N. and Wiersma, D. *Agron J* 1975, 67:777-781.
25. Mouat, M.C.H. and Nes, P. *Austr J Soil Res* 1986, 24:435-440.
26. Jones, C.A., Cole, C.V., Sharpley, A.N. et al. *Soil Sci Soc Am J* 1984, 48:800-805.
27. Barber, S.A., MacKay, A.D., Kuchenbuch, R.O. et al. *Plant Soil* 1988, 111:267-269.
28. Holdford, I.C.R. *Austr J Soil Res* 1979, 17:495-504.
29. Chapin, F.S. *Ann Rev Ecol Syst* 1980, 11:233-260.
30. Gehring, C.A. and Whitman, T.G. *Tree Physiol* 1994, 9:251-255.
31. Jackson, R.B. Soil heterogeneity and its exploitation by plants. PhD Dissertation, Utah State University, Logan, Utah, 1992.
32. Caldwell, M.M. *In:* Exploitation of Environmental Heterogeneity by Plants. Caldwell, M.M. and Pearcy, R.W. (eds), Academic Press, San Diego, 1994, pp. 325-347.
33. Sanders, I.R. and Fitter, A.H. *New Phytol* 1992, 120:525-533.
34. Hetrick, B.A.D., Wilson, G.W.T. and Harnett, D.C. *Can J Bot* 1989, 67:2608-2615.

35. Amiotti, N., Bravo, O. and Lageyre, L. *In: Proceedings XX Congreso Argentino de Ciencias del Suelo*, Salta, Argentina, 2006, p. 511.
36. Busso, C.A., Brevedan, R.E., Flemmer, A.C. et al. *In:* Plant Physiology and Plant Molecular Biology in the New Millennium. Hemantaranjan A (ed), Scientific Publishers, Varanasi, 2003, pp. 341-395.
37. Cano, E. *Pastizales naturales de La Pampa. Descripción de las especies más importantes.* Convenio AACREA-Provincia de La Pampa, Buenos Aires, Argentina, 1988.
38. Bóo, R.M. and Peláez, D.V. *Bol Soc Arg Bot* 1991, 27:135-141.
39. Turner, N.C. *In:* Proceedings of the International Conference of Measurements of Soil and Plant Water Status. Utah State University, Logan, Utah, 1987, 2:13-24.
40. Williams, T.E. and Baker, H.K. *J Brit Grassl Soc* 1957, 12:49-55.
41. Phillips, J.M. and Hayman, D.S. *Trans Brit Mycol Soc* 1970, 55:58-162.
42. Bray, R.H. and Kurtz, L.T. *Soil Sci* 1945, 59:39-45.
43. Busso, C.A., Briske, D.D. and Olalde-Portugal, V. *Oikos* 2001, 93:332-342.
44. Steel, R.G. and Torrie, J.H. Principles and procedures of statistics. McGraw-Hill Company, New York, 1981.
45. Zar, J.H. Biostatistical Analysis. Prentice Hall, New Jersey, 1999.
46. Becker, G.F., Busso, C.A. and Montani, T. *J Arid Environ* 1997, 35:233-250.
47. Busso, C.A. and Richards, J.H. *Acta Oecol* 1993, 14:3-15.
48. Busso, C.A. and Richards, J.H. *J Arid Environ* 1995, 29:239-251.
49. Asay, K.H. and Johnson, D.A. *Crop Sci.* 1990, 30:70-82.
50. Johnson, D.A., Asay, K.H., Tieszen, L.L. et al. *Crop Sci* 1990, 30:338-343.
51. Asay, K.H., Jensen, K.B., Waldron, B.L. et al. *Agron. J.* 2002, 94:1337-1343.
52. Becker, G.F., Busso, C.A., Montani, T. et al. *J Arid Environ* 1997, 35:251-268.
53. Flemmer, A.C., Busso, C.A., Fernández, O.A. et al. *Arid Land Res Manage* 2003, 17:139-152.
54. Becker, G.F., Busso, C.A., Montani, T. et al. *J Arid Environ* 1997, 35:269-283.
55. Saint Pierre, C. and Busso, C.A. *Phyton* 2006, 75:21-30.
56. Noyd, R.K., Pfleger, F.L. and Russelle, M.P. *New Phytol* 1995, 129:651-660.
57. Jasper, D.A., Mandal, R. and Osman, K.T. *Soil Biol Biochem* 1979, 11:501-505.
58. Mengel, K. and Kirkby, E.A. *Principles of Plant Nutrition.* International Potash Institute, Bern, 1982.
59. Mohammad, M.J., Pan, W.L. and Kennedy, A.C. *Mycorrhiza* 1998, 8:39-144.
60. Baylis, G.T.S. *Search* 1972, 3:257-258.
61. Hetrick, B.A.D., Kitt, D.G. and Wilson, G.T. *Can J Bot* 1988, 66:1376-1380.
62. Allen, E.B. and Allen, M.F. *New Phytol* 1986, 104:559-571.
63. Ferguson, W.S. *Can J Soil Sci* 1964, 44:180-187.
64. Eck, H.V. and Fanning, C. *Agron J* 1961, 53:335-339.
65. Olsen, S.R., Watanabe, F.S. and Danielson, R.E. *Soil Sci Soc Am J* 1961, 25:289-294.
66. Gutiérrez-Boem, F.H. and Thomas, G.W. *Agron J* 1998, 90:166-171.
67. Flemmer, A.C., Busso, C.A., Fernández, O.A. et al. *Can J Plant Sci* 2002, 82:539-547.
68. Moretto, A.S. and Distel, R.A. *Austral Ecol* 2002, 27:509-514.
69. Moretto, A.S. and Distel, R.A. *Appl Soil Ecol* 2001, 18:31-37.
70. Moretto, A.S. and Distel, R.A. *J Arid Environ* 2003, 55:503-514.

6

Induced Resistance in Plants and the Role of Arbuscular Mycorrhizal Fungi

Y. Ismail and M. Hijri

Université de Montréal, IRBV, 4101 Rue Sherbrooke Est
Montreal, QC, H1X 2B2, Canada

Introduction

Control of plant diseases has become a formidable task for the farmers in their efforts to produce crops with high yield and quality. Farmers worldwide have relied upon chemical control for the management of plant diseases. However, the heavy use of chemical pesticides often leads to deterioration of the agro-ecosystems, in addition to its being hazardous to the health of humans and animals. Present ecological concerns about frequent fungicide applications are serious and include worker safety, contamination of drainage water and consumer exposure to fungicide residues. However, the use of fungicides, earlier used to protect the crop from diseases, has also been drastically reduced to protect the environment. Therefore, alternative control measures are needed to fulfill grower's efforts to achieve pesticide-free production. During the last ten years, a great effort has been done in several countries to develop alternative and safe control methods to plant diseases. Alternatively, induced resistance of plant pathogens has increased and indeed offers many attractions as an alternative strategy. Plants can be induced locally and systemically to become more resistant to diseases through various biotic or abiotic stresses. The biological inducers include necrotizing pathogens, non-pathogens, root colonizing bacteria or arbuscular mycorrhizal fungi. Resistance-inducing chemicals that are able to induce broad resistance to disease offer an additional option for the farmer to complement genetic disease resistance and the use of fungicides. Therefore, in this review, authors discuss further information about induced resistance to plant diseases, to understand how plants integrate pathogen-induced signals into specific defense responses with focusing on the most important key components including salicylic acid (SA), jasmonic acid (JA) and ethylene (ET); explaining the biochemical changes associated with induced resistance; understanding the role of bacteria in the rhizosphere and arbuscular mycorrhiza fungi as bio-inducers mediated for induced resistance against pathogens; and the ability and strength of exogenous application of chemical inducers such as SA, K_2HPO_4 and their role for inducing resistance.

The Early Events in Induced Resistance

Plants have evolved complex recognition and response mechanisms to counter attack by pathogens. Disease occurs only when the pathogen is able to avoid early detection by the plant. Pathogens have to search for their hosts even by passive or active ways. In early events associated with plant-pathogen recognition, plants activate battery of defense responses including the oxidative burst, the hypersensitive response (HR), cell wall fortification, and defense-related protein synthesis [1-5]. When a plant and a pathogen come into contact, close communications occur between the two organisms [6, 7]. Pathogen activities focus on colonization of the host and utilization of its resources, while plants are adapted to detect the presence of pathogens and to respond with antimicrobial defenses and other stress responses. Plant and pathogen species are often highly co-evolved, meaning for example that standard plant barriers to microbial infection can be circumvented by particular pathogen species, but also that otherwise successful pathogens can be blocked by the unique adaptive responses of certain plants. As an infection plays out, the plant's metabolism often represents a shifting mixture of disease resistance responses and disease susceptibility responses. Interactions between plants and pathogens induce a series of plant defense responses [1], including calcium and other ion fluxes, activation of protein kinases, production of signaling compounds such as reactive oxygen intermediates (ROIs), salicylic acid (SA), nitric oxide (NO), ethylene (ETH) and jasmonic acid (JA) which cause activation of many downstream responses such as synthesis of antimicrobial proteins, phenolics, flavonoids, phytoalexin compounds, and activation of cell wall reinforcement responses.

Signal Transduction

Signal transduction is one of the most pathways in the plant-pathogen incompatibility. SA, JA and ET are the most key components of defense signal transduction.

Salicylic Acid

Salicylic acid (SA) is an important signal molecule in plant defense. The pharmacological properties of salicylate derivatives from willow bark were for centuries prized by ancient Greeks and American Indians, eventually leading to the invention and marketing of aspirin (acetylsalicylic acid), a mammalian cyclooxygenase inhibitor and the oldest, most widely used drug in history [8]. The biosynthesis of SA is from phenylalanine [9] and depending on the plant species, either from free benzoic acid, benzoyl glucose or *o*-coumaric acid as direct precursors [10]. In some bacteria, SA is synthesized from chorismate via isochorismate. The enzyme isochorismate synthase (ICS) and isochorismate pyruvate lyase (IPL) catalyze the two steps from chorismate to SA [11]. Most of these findings were supported by results obtained in tobacco, rice, cucumber and potato that showed an accumulation of radiolabelled ^{14}C-SA in tissue incubated with ^{14}C-phenylalanine [10]. Suppression of the expression of one pheneylalanine ammonia-lyase (PAL) gene in transgenic tobacco leads to a decrease in SA accumulation [12]. The identification of SA-deficient *Arabidopsis thaliana* mutant sid2 [13] and the subsequent localization of this mutation in a gene encoding a functional isochorismate synthase (ICS) showed some new light on the biosynthetic pathway of SA [14, 15]. Only minimal levels of ozone exposure [13] provided a strong support for isochorismate as a precursor of stress-induced SA accumulation in *A. thaliana* [16]. The biosynthetic pathway of SA in *A. thaliana* is therefore related to that described in bacteria where SA is synthesized from chorismate via the rate-limiting enzyme ICS and isochorismate pyruvate lyase (IPL) [17]. An increase in SA levels and increased resistance to tobacco mosaic virus was shown to take place in transgenic tobacco plants transformed with bacterial ICS and IPL, indicating that this pathway can operate in plants [18]. The discrepancy in the biosynthetic pathway of SA proposed for *A. thaliana* and other plants was often assumed to reflect differences among species [16]. To clarify the biosynthesis of SA in tobacco, ICS gene from *Nicotiana benthamiana* (*NbICS*) was cloned and used virus-induced gene silencing (VIGS) to suppress the expression of *NbICS* in *N. benthamiana*. This led to an effective disruption of the accumulation of SA and of phylloquinone (PHQ) in plants undergoing a biotic and abiotic stress indicating that the ICS enzyme is also required for SA biosynthesis in *N. benthamiana* [16].

Many investigators indicated that SA has important role in the activation of plant defense responses against biotic and abiotic stresses [16, 19, 20]. Plant resistance to biotrophic pathogens is classically thought to be mediated through SA signaling [21]. By contrast resistance to necrotrophic pathogens is controlled by jasmonic acid (JA) and ethylene (ET)-signaling pathways and genetically, in concern that SA and JA / ET defense pathways interact antagonistically [22, 23]. These signal components are integrated in a signaling network together in plant-pathogen interactions. This network exhibits the typical hallmarks found in integrated systems including negative and positive crosstalk interactions and specific responses to combinations of stimuli [24].

The involvement of SA as a signal molecule in local defense and in systemic acquired resistance (SAR) has been extensively studied [25, 26]. SA is also required for symptoms development [27]. Increases in the endogenous levels of SA and its conjugates in pathogen-inoculated plants coincide with the elevated expression of genes encoding the pathogenesis-related proteins (PR) and the activation of disease resistance [28]. Evidence for the role of SA in basal resistance came from the analysis of transgenic plants expressing the bacterial salicylate hydroxylase (*NahG*) gene [29]. Salicylate hydroxylase inactivates SA by converting it to catechol. NahG tobacco plants were found to be more susceptible to *Tobacco mosaic virus* (TMV), the bacterium *Pseudomonas syringae* pv. *tabaci*, and the fungus *Cercospora nicotianae* [30].

Jasmonic Acid (JA)

It has become clear in the last decade that jasmonic acid (JA) is a key regulator in the development, physiology, and defense of plants, the complexity of the signaling network in which JA evolves is just emerging [31, 32]. JA is involved in carbon partitioning [33], in mechano-transduction [34], and the ability of plants to synthesize and perceive JA is absolutely essential for the correct development and release of pollen in *Arabidopsis* [33-38]. There is also strong evidence supporting a central role of JA in plant defense. Exogenous JA powerfully regulates the expression of many defense genes in plants, and its *in vivo* production and perception seem to be of vital importance in mounting successful defense against insect attackers [39-41]. Together with ethylene, JA also plays a crucial role in defense against necrotrophic fungi [22, 23, 42-44] and in induced systemic resistance in response to nonpathogenic rhizobacteria [45].

The biosynthesis of JA occurs through the octadecanoid pathway [46, 47] and is initiated by the addition of molecular oxygen to linolenic acid (18:3) to form 13-hydroperoxylinolenic acid (13-HPOTrE). This fatty acid, hydroperoxide, is then dehydrated by allene oxide synthase (AOS) and cyclized by allene oxide cyclase (AOC) to the cyclopentenone 12-oxo-phytodienoic acid (OPDA). Although the chemical nature of OPDA allows four stereoenantiomers, the concerted action of AOS and AOC generate exclusively 9S,13S-OPDA [48], the precursor of active 3R,7S-JA [49]. The next step in the formation of JA is the reduction of the pentacyclic ring double bond in 9S,13S-OPDA by the enzyme OPDA reductase 3 (OPR3) [50] to the cyclopentanone 3-oxo-2(2'[Z]-pentenyl)-cyclopentane-1-octanoic acid (OPC:8). Other related enzymes such as OPR1 and OPR2, initially thought to be involved in the biosynthesis of active 3R,7S-JA, have almost shown no activity against 9S,13S-OPDA [50]. Finally, OPC:8 is shortened by three cycles of β-oxidation to yield JA. In parallel, hexadecatrienoic acid (16:3) can be metabolized to dinor oxo-phytodienoic acid (dnOPDA) [51], the 16-carbon structural homolog of OPDA. Although attention has focused mainly on JA as a signal, the possibility that the JA precursor OPDA could itself be biologically active has been proposed [34, 51-53]. However, there is also strong evidence, based on detailed pharmacological studies, that the natural oxylipin capable of eliciting the synthesis of alkaloids like sanguinarine in *Eschscholtzia* cell cultures is JA [54].

Ethylene (ET)

Ethylene is a gaseous plant hormone that affects myriad developmental processes and fitness responses, including germination, flower and leaf senescence, fruit ripening, leaf abscission, root nodulation, programmed cell death, and responsiveness to stress and pathogen attack [55, 56]. A well-known effect of ethylene on plant growth is the so-called 'triple response' of etiolated dicotyledonous seedlings. The role of ethylene in the hormonal regulation of plant development has been well established. In addition, it has been implicated in biotic stress, both as a virulence factor of fungal and bacterial pathogens and as a signaling compound in disease resistance [57]. Enhanced ethylene production is an early, active response of plants to perception of pathogen attack and is associated with the induction of defense reactions [58]. It is generally assumed that ethylene production during stress contributes to stress alleviation, but several plant pathogenic fungi and bacteria are capable of producing ethylene as a virulence factor, which improves their ability to colonize plant tissues [59, 60]. For instance, the ability of the bacterial leaf pathogen *Pseudomonas syringae* pv. *glycinea* to proliferate in the leaves of its host plant soybean is impaired in mutants that lack the capacity to produce ethylene [61]. Such observations indicate that ethylene produced during infection promotes disease rather than alleviates it. Indeed, ethylene is responsible for the epinasty and defoliation caused by the soil-borne fungus *Verticillium dahliae* in cotton [62], and for the stunting and chlorosis of cucumber infected by cucumber mosaic virus [63, 64].

Application of exogenous ethylene in the chemical form, ethephon (2-choloethylphosphonic acid), even as seed soaking or as foliar spray on faba bean plants had significant reduction of foliar fungal pathogens *Botrytis fabae*, *Uromyces vicia-fabae* and *Alternaria solani*. In addition, it had effect on plant yield of seeds [20]. The mode of action of ethylene as a modulator of disease resistance in plants seems to depend on the timing of the exposure of plant whether resistance is stimulated or reduced [57]. For example, the causal agent of gray mould, the fungus *Botrytis cinerea*, is able to infect a wide range of vegetables, ornamentals and fruits. Ethylene treatments typically promote disease development [65] but, on carrot, ethylene appears to be involved in resistance [66]. In general, treatments with ethylene promote leaf senescence and fruit ripening, which can make tissues either more susceptible to disease or more resistant [67]. Often, ethylene treatment must increase disease development simply through its acceleration of ripening or senescence. In addition, the conditions under which experiments have been carried out are not always clearly specified. Furthermore, various abiotic stresses can inadvertently affect plant susceptibility to disease. However, several observations indicate that when ethylene is applied before inoculation with a pathogen, it reduces or has no effect on disease development, whereas disease development is accelerated when plants are treated with ethylene after infection [55]. Thus, it seems that the timing of the exposure of plants to ethylene can determine whether resistance is stimulated or reduced (Figure 1).

Van Loon et al. [57] showed the involvement of ethylene in SAR. Ethylene can induce certain types of pathogenesis-related proteins or phytoalexins, and, through stimulation of the phenylpropanoid pathway, can rigidify cell walls in various plant species [55,59]. However, in several studies it has been demonstrated that ethylene is not necessary for these defensive activities to be expressed, or that comparable defenses are regulated differentially by ethylene, jasmonic acid or salicylic acid in different plant species. For example, the salicylic acid-dependent defense pathway is involved in resistance against *Botrytis cinerea* in tomato, but not in tobacco; whereas the salicylic acid dependent defense pathway is involved in resistance against the powdery

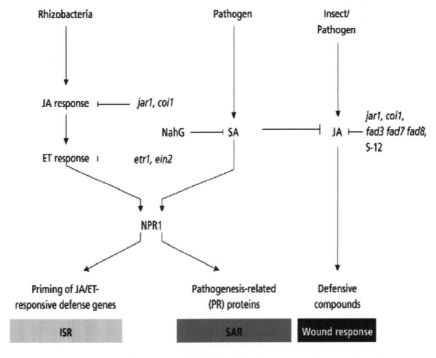

Figure 1: Signal transduction pathways leading to the induction of ISR, SAR and the wound response (adapted from [22]).

mildew fungus *Oidium neolycopersici* in tobacco but not in tomato [68]. In general, ethylene appears to stimulate and enhance defense responses [69, 70]. Treatment of *Arabidopsis* seedlings with 1 mM ACC enhances resistance against *P. syringae* pv. *tomato*; this induced resistance response strongly resembles the induced systemic resistance (ISR) that is elicited by specific strains of non-pathogenic, root colonizing bacteria, such as *Pseudomonas fluorescens* strain WCS417 [71]. ACC activates the PR-4 type hevein (Hel) gene, which can be used as a marker of this treatment [72]. However, when triggering ISR, WCS417 bacteria neither increase ethylene production nor activate ethylene-dependent PR gene expression [73]. Instead, ethylene responsiveness is required for ISR to be induced and expressed [74]. Hence, the ethylene response mutants etr1-1, ein2-1-ein7 and axr1-12 do not express ISR in response to root colonization by strain WCS417 [75], and neither does the enhanced disease susceptibility (eds) mutant eds4-1 nor the accessions RLD and Wassilewskija-0, which all have reduced responsiveness to ethylene and develop more severe symptoms after infection with *P. syringae* pv. *tomato* [73].

These results indicate that in Arabidopsis, ethylene is required for basal resistance against *P. syringae* pv. *tomato*. Moreover, ethylene perception is required for the plant to react to rhizobacteria by developing ISR. The eir1-1 mutant, which is insensitive to ethylene in the roots but not in the leaves, does not express ISR when WCS417 is applied to the roots, but does when the inducing bacteria are infiltrated into the leaves [75], which demonstrates that ethylene responsiveness is required at the site of resistance induction by ISR-eliciting rhizobacteria. Apparently, treatment with ACC can activate the same pathway leading to induced resistance and the ethylene mediates the generation of a mobile signal that enhances resistance systemically [45]. Induced systemic resistance (ISR) also, dependent on jasmonic acid responsiveness [57], studies on JA and ET indicated that treatment with 0.1 mM of methyl-jasmonate enhances resistance and activates the JA-responsive genes *Vsp2* and *Pdf1.2*, which encode a vegetative storage protein and plant defensin, respectively. Rhizobacteria-mediated ISR does not involve induction of these genes either but, upon challenge inoculation of induced plants with *P. syringae* pv. *tomato*, expression of jasmonic acid and ethylene-dependent genes is accelerated and enhanced [76]. Like ethylene non-responsiveness, a deficiency in jasmonic acid perception has been shown to increase the susceptibility of *Arabidopsis* and tomato to a wide range of pathogens with different lifestyles [77-79]. Yet, in the generation of ISR in *Arabidopsis*, eliciting enhanced protection by jasmonic acid is dependent on ethylene responsiveness because treatment with methyl jasmonate fails to elicit ISR in etr1 or ein2 mutants [74]. Ethylene and jasmonic acid cooperate in inducing ethylene response factor 1 (ERF1), which drives the activation of defense-related genes such as PR-4 and Pdf1.2 and positively regulates the expression of jasmonic acid-inducible genes involved in defense responses [80-82].

Biochemical Changes Associated with Induced Resistance

Resistance reactions taking place in the non-infected parts of the pretreated plants can be studied separately from reactions occurring at the infection site. At the site of attack, the resistance responses of the host includes modifications of the cell wall [83] production of phytoalexins [84], synthesis of pathogenesis related (PR) proteins [20, 57, 85-87], or activation of programmed cell death also called the hypersensitive reaction (HR) [88-91]. Infection of plants by viruses, fungi and bacteria frequently leads to the induction of pathogenesis-related (PR) proteins, which also may be induced by chemical treatments or by any stress factor [92]. The PR proteins such as chitosanase and peroxidase are present in any higher plants, and both enzymes have been implicated in defense reactions of plants against potential pathogens. Several PR protein families are identified and classified into PR17 (Table 1) [57]. The products of these enzymes are thought to act as elicitors for further induction of the enzymes and for activation of other defense biochemicals such as phytoalexins production and lignifications [93].

Pretreatments with salicylic acid (SA), a possible signal for SAR or functionally related inducers such as 2,6 dichloroisonicotinic acid (INA) or benzothiadiazole (BTH; BION), potentiate elicitor induced H_2O_2 generation

Table 1. Recognized families of pathogenesis-related proteins

Family	Type of member	Properties	Gene symbols
PR-1	Tobacco PR-1a	Unknown	Ypr1
PR-2	Tobacco PR-2	β-1, 3-glucanase	Ypr2 [Gns2 ('Glb')]
PR-3	Tobacco P, Q	Chitinase type I, II, IV, V, VI, VII	Ypr3, Chia
PR-4	Tobacco 'R'	Chitinase type I, II	Ypr4, Chid
PR-5	Tobacco S	Thaumatin-like	Ypr5
PR-6	Tomato Inhibitor I	Proteinase-inhibitor	Ypr6, Pis ('Pin')
PR-7	Tomato P69	Endoproteinase	Ypr7
PR-8	Cucumber chitinase	Chitinase type III	Ypr8, Chib
PR-9	Tobacco 'lignin-forming peroxidase'	Peroxidase	Ypr9, Prx
PR-10	Parsley 'PR1'	Ribonuclease-like	Ypr10
PR-11	Tobacco 'class V' chitinase	Chitinase, type I	Ypr11, Chic
PR-12	Radish Rs-AFP3	Defensin	Ypr12
PR-13	Arabidopsis THI2.1	Thionin	Ypr13, Thi
PR-14	Barley LTP4	Lipid-transfer protein	Ypr14, Ltp
PR-15	Barley OxOa (germin)	Oxalate oxidase	Ypr15
PR-16	Barley OxOLP	Oxalate-oxidase-like	Ypr16
PR-17	Tobacco PRp27	Unknown	Ypr17

Further details can be found at http://www.bio.uu.nl/~fytopath/PR-families.htm

or expression of defense related genes such as phenylalanine ammonia-lyase (PAL) or 4-coumarate:CoA ligase [20, 94-96]. For instance, SA, 2,6-dichloroisonicotinic acid (INA) and benzothiadiazole (BTH) induce the same set of PR genes that is induced upon biological induction of SAR. Moreover, their action often involves signaling steps that are also required for the expression of SAR [97, 98]. The search for a SA-binding protein has led to catalase and ascorbate peroxidase [99]. The binding of SA to such enzymes might lead to the formation of a phenolic radical that in turn is involved in lipid peroxidation. The products of lipid peroxidation can activate defense gene expression [100]. Treatment of potato plants with acetylsalicylic acid (ASA), 2,6-dichloroisonicotinic acid (INA), and benzothiadiazole (BTH), in order to induce systemic resistance against leaf and tuber diseases, resulted in increased resistance against Alternaria leaf spot and powdery mildew diseases. Levels of the enzyme beta-1,3-glucanase increased several times in the treated leaves and stem and by about 30% in tubers [101].

The correlation between resistance and the induction of enzymes of phenyl propanoid metabolism, involved for example in the syntheses of lignin and isoflavonid phytoalexin has become a model in the study of plant disease resistance. In plants, peroxidase is involved in many processes that lead to morphological changes associated with disease resistance [102]. In particular, peroxidase activity has been reported to be a biochemical marker for resistance and to be associated with systemic resistance [103]. However, peroxidase activity increased in inoculated leaves, so there is an active response in systemic induced resistance as well [20, 57, 104]. Activation of peroxidase is a general response of infected plants and tissue peroxidase level has been well correlated with resistance [105]. Abou El-Hawa [106] found that, foliar application with hydroxyphaseolin (HP) a natural antibiotic, phytoalexin, and was initially formed by soybean (*Glycin max*), could induce resistance against chocolate spot and gray mold diseases caused by *B. fabae* and *B. cinerea* in *Vicia faba*, and *Phaseolus vulgaris*, respectively. This finding was found to be attributed to the ability of HP to stimulate the formation of phytoalexins specific to the host wyerone acid in *Vicia faba* and phaseolin in *Phaseolus vulgaris* in sufficient quantities to resist the parasite. Kuti and Nawar [107] found that infection of broad bean leaves with *Botrytis fabae* resulted in rapid accumulation of the phytoalexin wyerone acid and increase in peroxidase activities in resistant and susceptible broad bean cultivars. Rapid accumulation of wyerone acid equal to or greater than 2-fold in leaves of resistant than susceptible broad bean cultivars was observed. Peroxidase activity in infected leaves of resistant cultivars was greater than 2-fold and less than 2-fold in leaves of susceptible cultivars when compared with uninfected leaves.

Induced Resistance by Rhizosphere Bacteria

Nonpathogenic rhizobacteria can induce a systemic resistance in plants that is phenotypically similar to pathogen-induced systemic acquired resistance (SAR)

[108]. Rhizobacteria-mediated induced systemic resistance (ISR) has been demonstrated against fungi, bacteria, and viruses in various plants, Arabidopsis, bean, carnation, cucumber, radish, tobacco, and tomato under conditions in which the inducing bacteria and the challenging pathogen remained spatially separated [108]. Induced resistance triggered by biological agents can be subdivided into two broad categories. The classical type of biologically induced resistance is often referred to as systemic acquired resistance (SAR), and occurs in distal plant parts after localized infection by a necrosis-inducing pathogen. Ross [109] was the first who provided a detailed description of the SAR phenomenon. He demonstrated that hypersensitively reacting tobacco developed enhanced resistance in non-inoculated leaves against subsequent infection by tobacco mosaic virus (TMV). The second type of biologically induced resistance develops systemically in response to colonization of plant roots by selected strains of non-pathogenic rhizobacteria. Two research groups independently demonstrated that rhizosphere-colonizing *Pseudomonas* spp. have the potential to enhance the resistance of the host plant [110, 111]. This type of induced resistance, generally called rhizobacteria-mediated induced systemic resistance [112], has been demonstrated in different plant species under conditions in which the rhizobacteria remained spatially separated from the challenging pathogen [108].

The common procedures to apply induced resistance by plant rhizobacteria are using a suspension of the bacteria on, or mixing it with, autoclaved soil; dipping seedling roots in a bacterial suspension at transplanting; or coating seeds with high numbers of bacteria before sowing [113]. Subsequently, seedlings are challenged with a pathogen. Because rhizobacteria are present on the roots, systemic protection against root pathogens must be demonstrated by applying the inducing bacteria to one part of the root system and the challenging pathogen to another part, for instance by making use of split-root systems [108]. Testing for protection against foliar pathogens is easier, because the pathogens are naturally separated from the rhizobacteria. However, rhizobacteria applied to seeds or to soil into which seeds are sown or seedlings are transplanted, can move into the interior of aerial plant tissues and maintain themselves to some extent on the exterior of aerial surfaces [114]. Examples for enhancing induced resistance in plants mediated by nonpathogenic rhizobacteria, treatment of bean seeds with *Pseudomonas flurescens* S97 reduced the number of lesions due to halo blight to 17% of that in nontreated controls [115]. Using *P. fluorescens* WCS417-mediated protection of carnation against Fusarium wilt [116], and resistance in cucumber against anthracnose induced by any of six PGPR [117] stems, petioles, cotyledons and/or leaf extracts were free from, or contained at most a negligible quantity of, inducing bacteria, implying involvement of induced systemic resistance (ISR).

Experiments were conduced over six years period to develop the role of PGPR-mediated induced resistance in cucumber against bacterial wilt caused by *Erwinia tracheiphila* [118]. The initial experiments investigated the factors involved in treatment with PGPR led to ISR to bacterial wilt disease in cucumber. Results demonstrated that PGPR-ISR against bacterial wilt and feeding by the cucumber beetle vectors of *E. trachiphiela* were associated with reduced concentrations of cucurbitacin, a secondary plant metabolite and powerful beetle feeding stimulant. In another experiment, PGPR induced resistance against bacterial wilt in the absence of the beetle vectors, suggesting that PGPR-ISR protects cucumber against bacterial wilt not only by reducing beetle feeding and transmission of the pathogen, but also through the induction of other plant defense mechanisms after the pathogen has been introduced into the plant. The mechanisms of resistance induced by nonpathogenic rhizobacteria are little complicated and they are more than one mechanism. Many rhizobacteria triggering ISR can inhibit growth of the pathogen directly by antagonistic effect or ISR can be an important consequence of tissue colonization with nonpathogenic rhizobacteria [119]. In order to prove that resistance-mediated by nonpathogenic rhizobacteria is induced systemically in plant tissue, it must be shown that inducing rhizobacteria are absent from the site of challenge with the pathogen and that the inducing bacterium and the challenging pathogen remain spatially separated for the duration of the experiment [108]. Whether the disease suppression resulted from antagonism or from ISR is not clear, because further experiments on absence or existence of the antagonistic bacteria from the aerial parts are needed. Meanwhile, injection of cucumber cotyledons *P. putida* 89B-27 and *Serratia marcescens* 90-166 multiplied in the tissue but were not recovered from stems 1 or 2 cm above or below the cotyledons [120]. Thus, while some bacteria that induce systemic resistance colonize internal tissues, they do not appear to establish themselves on challenged leaves, suggesting that neither competition nor antibiosis is involved in disease suppression [121].

Fluorescent *Pseudomonas* spp. have been studied for decades for their plant growth-promoting effects through effective suppression of soil-borne plant diseases by many investigators especially van Loon and his group in Utrecht University, Netherlands. He indicated that the modes of action that play a role in disease suppression by these bacteria include siderophore-mediated competition for iron, antibiosis, production of lytic enzymes, and induced systemic resistance (ISR). The involvement of ISR is typically studied in systems in which the *Pseudomonas* bacteria and the pathogen are inoculated and remain spatially separated on the plant, e.g., the bacteria on the root and the pathogen on the leaf, or by use of split root systems. Since no direct interactions are possible between the two populations, suppression of disease development has to be plant-mediated. The component of ISR-mediated by *Pseudomonas* bacteria is discussed by many investigators as flagella, lipopoly-saccharides and iron-regulated metabolites [121].

Flagella

Flagellins is the main protein component of flagella and can elicit defense responses in plants [122, 123]; the involvement of flagella in ISR by

Pseudomonas putida strain WCS358 was studied in *Arabidopsis*, bean, and tomato by applying isolated flagella and by using non-motile mutants that lack flagella [124]. In *Arabidopsis*, application of WCS358 flagella triggered ISR against *P. syringae* pv. *tomato*, whereas in bean or tomato, their application did not lead to induced resistance. Moreover, a mutant of WCS358 that lacks flagella was as effective in triggering ISR in *Arabidopsis* as the parental strain, and it was concluded that there are additional determinants in strain WCS358 that can induce resistance. These results suggest that flagella can be involved in ISR, with other components.

Lipopolysaccharides

Lipopolysaccharides (LPS) involved in cell components of pathogenic bacteria can induce ISR [121, 124, 125]; *P. fluorescens* WCS417r was used against *Fusarium oxysporum* f. sp. *dianthi* in carnation [126]. Bacterial cells were killed and LPS were purified of *P. fluorescens* WCS417r and applied to the carnation roots which subsequently triggered ISR to a level similar to that obtained by treatment with viable bacterial cells. Accumulation of phytoalexins after challenge inoculation of carnation treated with either viable or heat-killed cells or with LPS of WCS417 was significantly increased compared with that in control plants that were challenged with *Fusarium* spp. [126].

In radish, LPS was also demonstrated to be of importance in ISR by *P. fluorescens* strains WCS374r and WCS417r against wilt caused by *F. oxysporum* f. sp. *raphani* [127]. For evidence that the LPS have involvement in ISR was obtained from the use of mutants of the strains that lack the 0-antigenic side chain of the LPS. The 0-antigen minus mutants did not reduce disease incidence, whereas application of LPS reduced Fusarium wilt of radish to a level comparable to that after treatment with the wild type bacteria. In *A. thaliana*, LPS of *P. fluorescens* WCS417r and *P. putida* WCS358 appears to be involved in ISR against *P. syringae* pv. *tomato*, since applying isolated LPS triggered ISR. However, in this system, mutants lacking the 0-antigen were as effective as the parental strain, suggesting redundancy in ISR triggering traits in these strains [124, 128, 129].

Competition for Iron and the Role of Siderophores

Iron is an essential growth element for all living organisms. Under iron-limiting conditions PGPB including fluorescent pseudomonas produce low-molecular-weight compounds called siderophores to competitively acquire ferric ion [121, 130, 131]. Although various bacterial siderophores differ in their abilities to sequester iron, in general, they deprive pathogenic fungi of this essential element since the fungal siderophores have lower affinity [132,133]. Some PGPB strains go one step further and draw iron from heterologous siderophores produced by cohabiting microorganisms [130, 131, 134-136]. Siderophore biosynthesis is generally tightly regulated by iron-sensitive Fur proteins, the global regulators,

GacS and GacA, the sigma factors RpoS, PvdS, and FpvI, quorum-sensing autoinducers such as N-acyl homoserine lactone, and site-specific recombinases [137, 138]. However, some data demonstrate that none of these global regulators is involved in siderophore production. Neither GacS nor RpoS significantly affected the level of siderophores synthesized by *Enterobacter cloacae* CAL2 and UW4 [139]. RpoS is not involved in the regulation of siderophore production by *Pseudomonas putida* strain WCS358 [140]. In addition, GrrA/ GrrS, but not GacS/GacA, are involved in siderophore synthesis regulation in *Serratia plymuthica* strain IC1270, suggesting that gene evolution occurred in the siderophore-producing bacteria [141]. A myriad of environmental factors can also modulate siderophores synthesis, including pH, the level of iron and the form of iron ions, the presence of other trace elements, and an adequate supply of carbon, nitrogen, and phosphorus [142].

Arbuscular Mycorrhizal (AM) Fungi-Mediated Induced Resistance

Arbuscular mycorrhizal fungi (AMF) form symbiotic association with the root systems of most agricultural, horticultural and hardwood crop species; thus they are widespread potential biocontrol agents. AMF-symbiosis can reduce root diseases caused by several soil-borne pathogens [143]. AMF symbioses have a significant impact on plant interactions with other organisms. Increased resistance to soil-borne pathogens has been widely described in mycorrhizal plants. During mycorrhiza formation, modulation of plant defense responses occurs, potentially through cross-talk between salicylic acid and jasmonate dependent signaling pathways. This modulation may impact plant responses to potential enemies by priming the tissues for a more efficient activation of defense mechanisms [144]. In fact, AMF symbiosis confers numerous benefits to host plants, including improved plant growth and mineral nutrition [145, 146], tolerance to disease [147-149], and abiotic stresses such as drought [150], chilling [151] and salinity [152].

Alleviation of damage caused by soil-borne pathogens has been widely reported in AMF plants. Most studies on protection by AMF deal with the reduction of incidence and/or severity of soil-borne diseases—mainly root rot or wilting caused by fungi such as *Rhizoctonia*, *Fusarium*, or *Verticillium* and root rot caused by oomycetes including *Phytophthora*, *Pythium*, and *Aphanomyces* [143, 144]. Studies about AMF effects on foliar diseases are few and less conclusive. AM symbioses have been associated with enhanced susceptibility to biotrophic pathogens including viruses [153], powdery mildew, and rust fungi (*Blumeria*, *Oidium* and *Uromyces*), although increased tolerance was often observed in terms of plant mass and yield [154, 155]. Mycorrhization, however, reduced disease symptoms caused by a phytoplasma, and protection against the necrotroph; *Alternaria solani* has been shown in mycorrhizal tomatoes [156, 157]. Only recently, the effect of AMF on plant interactions with shoot pathogenic bacteria has been assessed: mycorrhizal symbiosis results in enhanced resistance to *Xanthomonas campestris* in *Medicago truncatula* [158].

Induced resistance-mediated by AMF was studied on *Phytophthora* infection on tomato plants [143], the ability of two AMF (*Glomus intraradices* and *Glomus mosseae*) to induce local or systemic resistance to *P. parasitica* in tomato roots. *G. mossae* was effective in reducing disease symptoms produced *P. parasitica* infection. The biochemical analysis to determine a local induction of mycorrhiza-related new isoforms of the hydrolytic enzymes chitinase, chitosanase, and β-1,3-glucanase. The super oxide dismutase is an enzyme involved in cell protection against oxidative stress. The activities of lytic enzymes of AMF-tomato plants were enhanced against *Phytophthora* cell wall of root protein extracts which indicated systemic effects of mycorrhizal symbiosis on tomato resistance to *Phytophthora*.

Pathogenesis related (PR) proteins were strongly expressed and monitored in leaves of tobacco (*Nicotiana tabacum* cv. Xanthinc) plants colonized by the arbuscular mycorrhizal fungus *Glomus intraradices,* against infection *with* the leaf pathogens *Botrytis cinerea* or tobacco mosaic virus. Infected leaves of mycorrhizal plants showed a higher incidence and severity of necrotic lesions than those of nonmycorrhizal controls. The similar responses were obtained in plants treated with low (0.1 mM) and high (1.0 mM) nutritional P levels and with mutant plants (NahG) that are unable to accumulate salicylic acid. Application of PR-protein activators BTH (benzothiadiazole) and INA (2,6-Dichloroisonicotinic acid) induced PR-1 and PR-3 expression in leaves of both nonmycorrhizal and mycorrhizal plants [159].

Histopathological studies indicate cell defense responses induced by arbuscular mycorrhizal fungus *Glomus mosseae* against *Phytopthora parasitica* in tomato roots [160]. The localized and systemic resistance was observed in AM-plants. The localized resistance of mycorrhizal tissues is observed in the rare occurrence of *P. parasitica* hyphae and their inability to penetrate AM-containing plant cells. The ISR is characterized by large reductions in root damage and in *P. parasitica* development inside nonmycorrhizal root tissues of mycorrhizal systems, in comparison to roots of nonmycorrhizal plants. This bioprotection is directly linked to root colonization by the arbuscular mycorrhizal fungus since neither a microbial filtrate from the *G. mosseae* inoculum nor low mycorrhizal levels are sufficient to induce it. The present cytomolecular investigations provide clear evidence that both the localized and the systemic protective effects induced by arbuscular mycorrhiza involve the accumulation of plant defense-related molecules in association with the elicitation of wall reactions in the host roots. This study by Cordier et al. [160] indicate both physical and chemical resistance involved in AMF-mediated ISR. The resistance of arbuscule-containing cortical cells in mycorrhizal root tissues is characterized by the accumulation of fluorescent compounds and the induction of pectin-free wall appositions that are rich in callose. It is well known that callose reinforces wall barriers produced early during plant defense reactions [161, 162]. The systemically induced resistance in nonmycorrhizal root parts is characterized by elicitation of host wall thickenings containing non-esterified pectins and PR-1a protein in reaction to intercellular pathogen hyphae, and by the formation of callose-rich encasement material around *P. parasitica* hyphae that are penetrating root cells. PR-1a protein is detected in the pathogen wall only in these tissues. None of these cell reactions are observed in nonmycorrhizal pathogen-infected root systems, where disease development leads to host cell death.

Induced Resistance in Plants by Chemicals

There are numerous reports demonstrating that resistance can be systemically induced in a number of plants by prior treatment with simple chemical substrates [20, 163-166]. Localized applications of various salts, such as phosphates, silicates and oxalates, to plants are reported to systemically induce resistance to a range of pathogens [167, 168]. Activation was usually highest when there was lesion like tissue damage at the points of application, suggesting that these chemicals mimic the biological SAR induction by inducing local lesion. Local lesion formation may also be the initial step for SAR type's resistance induction by unsaturated fatty acids in potato [169] and that described for harpin proteins from *Phytophthora* species [170].

Application of mono- or di-potassium phosphate salts as foliar spray treatments controlled powdery mildew of cucumber [170, 171], powdery mildew of sugar beet [172], rice blast disease [173], and foliar fungal diseases of faba bean [20]. Meanwhile, Reuveni and Reuveni [174] indicated that foliar sprays of phosphate and potassium salts can induce systemic protection against foliar pathogens in various crops such as cucumber, maize, rose, grapevine, apple, mango and nectarine. Mucharromah and Kúc [167] found that spraying aqueous solutions of oxalate, potassium phosphate dibasic or tribasic on the upper surface of leaf No. 1 of cucumber plants, induced systemic resistance to *Colletotrichum lagenarium*, *Cladosporium cucumerinus* and *Dydimella bryoniae*. Mosa [172] studied the potential of various abiotic and biotic agents to induce systemic resistance (ISR) in rice against blast disease, significant differences in the severity of blast disease on rice leaves among different inducers and method of application were recorded. The great reduction of leaf and panicle blast severity was obtained with BTH, when sprayed twice at both growth stages.

Various chemicals have been discovered that seem to act at various points of the defense activating network [166]. Some compounds e.g. polyacrylic acid and salicylic acid, and a lignification-inducing factor (LIF) has been shown to induce resistance in plants [174, 175]. Mills and Wood [175] found that pretreatment of cucumber plants with salicylic, acetylsalicylic or polyacrylic acid induced local and to a lesser extent systemic resistance to subsequent infection with *Colletorichum lagenarium*. Resistance is effectively induced by chemicals including benzoic acid derivatives such as salicylic acid (SA) and also by ethephon when sprayed on or injected into leaves, or watered into soil [175]. Although these compounds appeared to have little direct effect on pathogens, when coupled with the host's natural defense mechanisms, they might provide the competitive edge required to reduce disease [176, 177].

Ethephon seed treatment could induce resistance in faba bean plants against chocolate spot and Stemphylium blight [178]. Abou El-Hawa [179] found that foliar application with hydroxyphaseolin (HP) a natural antibiotic, phytoalexin, initially formed by soybean, could induce resistance against chocolate spot and gray mould diseases caused by *B. fabae* and *B. cinerea* in *Vicia faba*, and *Phaseolus vulgaris*, respectively. Application of K_2HPO_4 and $FeCl_3$ as foliar spray reduced browning and spreading of lesions caused by *Botrytis fabae* on detached leaves of faba bean [20].

Recently, two classes of chemicals were discovered that mimic the biological activation of SAR by necrotic pathogen, 2,6-dichloro isonicotinic acid (INA) and its derivatives and the benzo(1,2,3) thiadiazole derivatives [180] with s-methyl benzo (1,2,3) thiadiazole-7-carbothiate (acibenzolar-s-methyl) as the first commercial product marketed under the trade names BION, ACTIGARD and BOOST. These chemicals do not show any antimicrobial activity, *in vitro*, and activate resistance against the same spectra of pathogens as the biological inducers of SAR on the plant species where this information is available [166].

Chemically induced resistance has been intensively studied in various plant-pathogen systems using both biotic and abiotic inducing agents [163]. How these chemicals act to restrict disease development is still unknown. In the experiments with K_2HPO_4, protection was effective against three different fungal diseases of faba bean. KH_2PO_4 and K_2HPO_4 have been shown to induce systemic resistance in cucumber against *Colletorichum lagenarium* [164] and *Sphaerotheca fuliginia* [165, 171] in grapevine against *Uncinula necator* [165], and in faba bean against foliar fungal pathogens [20]. These studies showed high levels of protection by potassium phosphates against the respective diseases. It has been suggested that, in this case, protection is triggered by a process involving the sequestering of calcium from host tissue [164, 181]. It is possible that these chemical agents elicit the release of a signal and trigger the plants general response to stress [167]. Many plant enzymes are involved in defense reactions against plant pathogens. These include oxidative enzymes such as peroxidase and polyphenol oxidase which catalyse the formation of lignin and other oxidative phenols that contribute to the formation of defense barriers for reinforcing the cell structure [182, 183]. K_2HPO_4 is an inducer of peroxidase and chitinase of both the local and systemic level [104]. However, Wicks et al. [184] have clearly demonstrated that phosphorus (phosphonic) acid (H_3PO_3) reduced incidence of *Plasmopara viticola* on grapes when applied as post-infection curative treatment, and gave better protection than the fungicide, metalaxyl. The phosphonate is the hydrolysis product of H_3PO_3 in plants and extremely mobile within plant tissues; it moves from leaf to leaf, both upward and downward.

Integration of Induced Resistance with other Control Methods

Chemical activation of induced resistance is not the new silver bullet against plant diseases, any more so than was the case for genetic resistance or other biological or chemical products for disease control. The activation of induced resistance on most crops is best used in combination with other methods of disease control, including genetic disease resistance of all types and sound crop management that can provide additional reduction of the disease pressure [185]. Where less resistant cultivars are preferred for yield, quality, or agronomic reasons, resistance activators can stimulate the plants to better protect themselves against some pathogens. Against other pathogens and where the level of genetic and activated resistance is not sufficient, fungicides or, where available, biological products can help assure healthy crops and high produce and food quality [185]. In some cases, mixtures of activators with reduced rates of appropriate fungicides have given excellent disease control, with the fungicide providing curative and short-term protection and activated resistance providing long-term protection [185, 186]. In other cases, activators can lower the fungicide load per season, thereby reducing the selection pressure for resistance against modern selective fungicides. Little information is available for the integration of biological disease control methods with induced resistance and with fungicides, except for copper and sulphur.

Research on biological and chemical activation of disease and insect resistance has taught us that plants possess complex networks of inducible defense pathways that can interact with each other. Of these pathways, the SA dependent SAR pathway seems to be the most robust to be exploited for practical crop protection. With the increasing set of *Arabidopsis* defense pathway mutants it will be much easier in the future to determine if and how novel chemical disease control agents interact with the plant's own defense network. This information will help the optimal utilization of the various signaling pathways for practical crop protection through novel genetic, biological, or chemical solutions. Unfortunately, no such model system is available yet in monocots, where much less is known about the signaling pathways involved in biological or chemical stimulation of disease resistance. However, the experience with the chemical plant activators available so far suggests that some basic inducible broad-spectrum defense responses are conserved across the plant kingdom [185].

Chemical activation of disease resistance in plants represents an additional option to maintain healthy crops and to prevent losses due to plant diseases. Against some pathogens, like bacteria and viruses, it may be the only chemical control option where genetic resistance is not sufficient. Against dynamic fungal pathogens with a history of resistance to fungicides or of adaptation to resistant cultivars, resistance activators can help prevent the emergence of adapted pathogen populations. Sustainable crop production systems need all methods available to manage plant health, so that in each case the growers, together with their customers, can make the right choice by weighing costs, benefits and risks for their specific cropping situation. Growers then will have one more tool and sustainable disease control has a greater chance, where resistance activators become available.

Conclusion

This review discusses further knowledge about the early events in the incompatible plant-pathogen interactions leading to induced resistance and the most important key components of signal molecules including SA, JA and ET in the plants upon attack by a range of pathogens. We showed the biosynthesis of each signal molecule and its role in inducing resistance. Despite knowledge about the biochemical changes associated with induced resistance and focusing on PR proteins and other antimicrobial compounds, we discussed also, the ability of inducing resistance biologically by rhizosphere bacteria and arbuscular mycorrhiza fungi. The plant growth promoting rhizobacteria *Pseudomonas flurecsence* has an important role in mediating plant defense responses against pathogen attacks and some studies showed that AMF (*Glomus intraradicis* and *G. mossae*) have promising effects for enhancing plant resistance. Moreover, enhancement of resistance against plant diseases by chemical inducers was reported by many investigators including exogenous application of SA, K_2HPO_4, BION and 2,6-dichloro isonicotinic acid (INA).

Acknowledgements

The authors wish to express their sincere appreciation and gratitude to Rachid Lahalli for generous assistance.

References

1. Hammond-Kosack, K.E. and Jones, J.D.G. *Plant Cell* 1996, 8:1773-1791.
2. Lamb, C.J. and Dixon, R.A. *Annu Rev Plant Physiol Plant Mol Biol* 1997, 48:251-275.
3. Bolwell, G.P. and Wojtaszek, P. *Physiol. Mol Plant Pathol* 1997, 51:347-366.
4. Wojtaszek, P. *Biochem J* 1997, 322:4158-4163.
5. Jones, J.D.G. and Dangl, J.L. *Nature* 2006, 444:323-329.
6. Hammond-Kosack, K.E. and Parker, J.E. *Curr Opin Biochem* 2003, 14:177-193.

7. Dangl, J.L. and Jones, J.D.G. *Nature* 2001, 411:826-833.
8. Weissmann, G. *Sci Am* 1991, 264:84-90.
9. Strawn, M.A., Marr, S.K., Inoue, K. et al. *J Bio Chem* 2007, 282:5919-5933.
10. Garcion, C. and Métraux, J.P. *In:* Plant Hormone Signalling. Hedden, P. and Thomas, S. (eds), Oxford, Blackwell Publishing Ltd, 2006, pp. 229-255.
11. Serino, L., Reimmann, C., Baur, H. et al. *Mol Gen Genet* 1995, 249:217-228.
12. Pallas, J.A., Paiva, N.L., Lamb, C. et al. *Plant J* 1996, 10:281-293.
13. Nawrath, C. and Métraux, J.P. *Plant Cell* 1999, 11:1393-1404.
14. Strawn, M.A., Marr, S.K., Inoue, K. et al. *J. Biol. Chem* 2007, 282:5919-5933.
15. Wildermuth, M.C., Dewdney, J., Wu, G. et al. *Nature* 2001, 414:562-565.
16. Catinot, J., Buchala, A., Abou-Mansour, E. et al. *FEBS Lett* 2008, 20:473-478.
17. Verberne, M.C., Bludi Muljono, A.B. and Verpoorte, R. *In:* Biochemistry and Molecular Biology of Plant Hormones. Libbenga, K., Hall, M. and Hooykaas, P.J.J. (eds), Elsevier, London, 1999, pp. 295-312.
18. Verberne, M.C., Verpoorte, R., Bol, J.F. et al. *Nature Biotech* 2000, 18:779-783.
19. Van Wees, S.C.M., De Swart, E.A.M., Van Pelt, J.A. et al. *Proc Natl Acad Sci* 2000, 97:8711-8716.
20. Ismail, Y.M., Soliman, G.I., Mosa, A.A. et al. *J. Agric. Sci. Mansoura Univ* 2004, 29:1205-1214.
21. Loake, G. and Grant, M. *Curr Opin Plant Biol* 2007, 10:466-472.
22. Pieterse, C.M.J. and Van Loon, L.C. *Trends Plant Sci* 1999, 4:52-58.
23. Glazebrook, J. *Annu Rev Phytopathol* 2005, 43:205-227.
24. Genoud, T., Santa Cruz, M.B.T. and Métraux, J.P. *Plant Physiol* 2001, 126:1430-1437.
25. Raskin, I. *Ann Rev Plant Physiol Plant Molec Biol* 1992, 43:439-463.
26. Dempsey, D.A., Shah, J. and Klessig, D.F. *Crit Rev Plant Sci* 1999, 18:547-575.
27. O'Donnell, P.J., Schmelz, E.Z., Moussatche, P. et al. *Plant J* 2003, 33:245-257.
28. Gaffney, T., Friedrich, L., Vernooij, B. et al. *Science* 1993, 261:754-756.
29. Raridan, G.J. and Delaney, T.P. *Genetics* 2002, 161:803-811.
30. Delaney, T.P., Uknes, S., Vernooij, B. et al. *Science* 1994, 266:1247-1250.
31. Stintzi, A., Weber, H., Reymond, P. et al. *Proc Natl Acad Sci USA* 2001, 98:12837-12842.
32. Schenk, P.M., Kazan, K., Wilson, I. et al. *Proc. Natl. Acad. Sci. USA* 2000, 97:11655-11660.
33. Mason, H.S. and Mullet, J.E. *Plant Cell* 1990, 2:569-579.
34. Weiler, E.W., Albrecht, T., Groth, B. et al. *Phytochemistry* 1993, 32:591-600.
35. Feys, B.J.F., Benedetti, C.E., Penfold, C.N. et al. *Plant Cell* 1994, 6:751-759.
36. McConn, M. and Browse, J. *Plant Cell* 1996, 8:403-416.
37. Sanders, P.M., Lee, P.Y., Biesgen, C. et al. *Plant Cell* 2000, 12:1041-1061.
38. Stintzi, A. and Browse, J. *Proc Natl Acad Sci USA* 2000, 97:10625-10630.
39. Howe, G.A., Lightner, J, Browse, J. et al. *Plant Cell* 1996, 8:2067-2077.
40. McConn, M., Creelman, R.A., Bell, E. et al. *Proc Natl Acad Sci USA* 1997, 94:5473-5477.
41. Walling, L.L. *J. Plant Growth Reg* 2000, 19:195-216.
42. Vijayan, P., Schokey, J., Lévesque, C.A. et al. *Proc. Natl. Acad. Sci. USA* 1998, 95:7209-7214.
43. Staswick, P.E., Yuen, G.Y. and Lehman, C.C. *Plant J* 1998, 15:747-754.

44. Penninckx, I.A., Thomma, B.P., Buchala, A. et al. *Plant Cell* 1998, 10:2103-2113.
45. Van Wees, S.C.M., de Swart, E.A.M., van Pelt, J.A. et al. *Proc Natl Acad Sci USA* 2000, 97:8711-8716.
46. Moons, A., Prinsen, A., Bauw, G. et al. *Plant Cell* 1997, 9:2243-2259.
47. Schaller, F. *J. Exp. Bot 2001,* 52:11-23.
48. Ziegler, J., Stenzel, I., Hause, B. et al. *J. Biol. Chem* 2000, 25:19132-19138.
49. Creelman, R.A. and Mullet, J.E. *Annu Rev Plant Physiol Plant Mol Biol* 1997, 48:355-387.
50. Schaller, F., Biesgen, C., Muessig, C. et al. *Planta* 2000, 210:979-984.
51. Weber, H., Vick, B.A. and Farmer, E.E. *Proc. Natl. Acad. Sci. USA* 1997, 94:10473-10478.
52. Weiler, E.W., Kutchan, T.M., Gorba, T. et al. *FEBS Lett* 1994, 345:9-13.
53. Dittrich, H., Kutchan, T.M. and Zenk, M.H. *FEBS Lett* 1992, 309:33-36.
54. Weiler, E.W., Kutchan, T.M., Gorba, T. et al. *FEBS Lett* 1994, 345:9-13.
55. Ecker, J.R. *Science* 1995, 268:667-675.
56. Johnson, P.R. and Ecker, J.R. *Annu. Rev. Genet* 1998, 32:227-254.
57. Van Loon, L.C., Geraats, B.P.J. and Linthorst, H.J.M. *Trends Plant Sci* 2006, 11:184-191.
58. Boller, T. *In:* The Plant Hormone Ethylene. Mattoo, A.K. and Suttle, J.C. (eds), CRC Press, 1991, pp. 293-314.
59. Dong, X. and Sa, J.A. *Curr Opin Plant Biol* 1998, 1:316-323.
60. Scofield, S.R., Tobias, C.M., Rathjen, J.P. et al. *Science* 1996, 274:2063-2065.
61. Weingart, H., Ullrich, H., Geider, K. et al. *Phytopathol* 2001, 91:511-518.
62. Tzeng, D.D. and De Vay, J.E. *Physiol. Plant Pathol* 1985, 26:57-72.
63. Marco, S. and Levy, D. *Physiol Plant Pathol* 1979, 14:235-244.
64. Willem, F.B., Stijn, L.D., Miguel, F.C. et al. *Annu Rev Phytopathol* 2006, 44:393-416.
65. Elad, Y. *Neth J Plant Pathol* 1993, 99:105-113.
66. Hoffman, R., Roebroeck, E. and Heale, J.B. *Physiol Plant* 1988, 73:71-76.
67. Panter, S.N. and Jones, D.A. *Adv Bot Res* 2002, 38:251-280.
68. Achuo, E.A., Audenaert, K., Meziane, H. et al. *Plant Pathol* 2004, 53:65-72.
69. Lawton, K.A., Potter, S.L., Uknes, S. et al. *Plant Cell* 1994, 6: 581-588.
70. Ohtsubo, N., Mitsuhara, I., Koga, M. et al. *Plant Cell Physiol* 1999, 40: 808-817.
71. Van Loon, L.C. and Bakker P.A.H.M. *In:* PGPR: Biocontrol and Biofertilization. Siddiqui, Z.A. (ed.), Springer, 2005, pp. 39-66.
72. Van Wees, S.C.M., Luijendijk, M., Smoorenburg, I. et al. *Plant Mol Biol* 1999, 41:537-549.
73. Ton, J., De Vos, M., Robben, C. et al. *Plant J* 2002, 29:11-21.
74. Pieterse, C.M.J., van Wees, S.C.M., van Pelt, J.A. et al. *Plant Cell* 1998, 10:1571–1580.
75. Knoester, M., Pieterse, C.M.J., Bol, J.F. et al. *Mol. Plant-Microbe Interact* 1999, 12:720-727.
76. Verhagen, B.W.M., Glazebrook, J., Zhu, T. et al. *Mol Plant–Microbe Interact* 2004, 17:895-908.
77. Thomma, B.P.H.J., Penninckx, I.A.M.A., Broekaert, W.F. et al. *Curr Opin Immunol* 2001,13:63-68.
78. Thaler, J.S., Owen, B. and Higgins, V. *Plant Physiol 2004*, 135:530-538.

79. Pozo, M.J., Van Loon, L.C. and Pieterse, C.M.J. *J Plant Growth Regul* 2004, 23:211-222.
80. Berrocal-Lobo, M., Molina, A. and Solano, R. *Plant J* 2002. 29:23-32.
81. Lorenzo, O., Piqueras, R., Sánchez-Serrano, J. et al. *Plant Cell* 2003, 15:165-178.
82. Lorenzo, O., Chico, J.M., Sanchez-Serrano, J.J. et al. *Plant Cell* 2004, 16:1938-1950.
83. Hammerschmidt, R. *Physiol Mol Plant Pathol* 1999, 55:77-84.
84. Hammerschmidt, R. *Annu Rev Phytopathol* 1999, 37:285-306.
85. Hunt, M.D. and Ryals, J.A. *Crit Rev Plant Sci* 1996, 15:583-606.
86. Van Loon, L.C. *Euro J Plant Pathol* 1997, 103:753-765.
87. Van Loon, L.C. and Van Strien, E.A. *Physiol Mol Plant Pathol* 1999, 55:85-97.
88. Gilchrist, D.G. *Annu Rev Phytopatol* 1998, 36:393-414.
89. Lamb, C. and Dixon, R.A. *Annu Rev Plant Physiol* 1997, 48:251-275.
90. Richfield, M.H., Aviv, D.H. and Dangl, J.L. *Curr Opinion Plant Biol* 1998, 1:480-485.
91. Grant, M. and Mansfield, J. *Curr Opin Plant Biol* 1999, 2:312-319.
92. Benhamou, N. *Trends Plant Sci* 1996, 1:233-240.
93. Mauch, F. and Staehelin, A.L. *Plant Cell* 1989, 1:447-457.
94. Katz, V.A., Thulke, O.U. and Conrath, U. *Plant Physiol* 1998, 117:1333-1339.
95. Thulke, O. and Conrath U. *Plant J* 1998, 14:35-42.
96. Kauss, H., Fauth, M., Merten, A. et al. *Plant Physiol* 1999, 120:1175-1182.
97. Ward, E.R., Uknes, J.S., Williams, C.S. et al. *Plant Cell* 1991, 3:1085-1094.
98. Uknes, S., Mauch-Mani, B., Moyer, M. et al. *Plant Cell* 1992, 4:645.
99. Durner, J. and Klessig, D.F. *Proc Natl Acad Sci USA* 1995, 92:11312-11316.
100. Farmer, E.E., Weber, H. and Vollenweider, S. *Planta* 1998, 206:167-174.
101. Bokshi, A., Morris, S., Deverall, B. et al. Induction of systemic resistance of potato. Australian Potato Research, Development and Technology Transfer Conference, 31 July to 2 August, 2000. Adelaide, Australia.
102. Dalisay, R.F. and Kuc, A.J. *Physiol Mol Plant Pathol* 1995, 47:329-338.
103. Schaffrath, U., Scheinpflug, H. and Reisener, H.J. *Physiol Mol Plant Pathol* 1995, 46:293-307.
104. Irving, H.R. and Kuæ, J. *Physiol Plant Pathol* 1990, 37:355-366.
105. Sridhar, R. and Ou, S.H. *Phytopathol Z* 1974, 79:222-230.
106. Abou El-Hawa, M. *Egypt J Microbiol* 1998, 33:147-154.
107. Kuti, O.J. and Nawar, H.F. *Phytopathol* 2003, 93:848.
108. Van Loon, L.C., Bakker, P.A.H. and Pieterse, C.M.J. *Annu Rev. Phytopathol* 1998, 36:453-483.
109. Ross, A.F. *Virol* 1961, 14:340-358.
110. Van Peer, R., Niemann G.J. and Schippers, B. *Phytopathol* 1991, 91:728-734.
111. Wei, G., Kloepper, J.W. and Tuzun, S. *Phytopathol* 1996, 86:221-224.
112. Pieterse, C.M.J., Van Wees, S.C.M., Hoffland, E. et al. *Plant Cell* 1996, 8:1225-1237.
113. Kloepper, J.W. *Bioscience* 1996, 46:406-409.
114. Lamb, T.G., Tonkyn, D.W. and Kluepfel, D.A. *Can. J. Microbiol* 1996, 42:1112-1120.
115. Alstrom, S.I. *J. Gen. Appl. Microbiol.* 991 37:495-501.
116. Van Peer, R., Niemann, G.J. and Schippers, B. *Phytopathol* 1991, 81:728-734.

117. Wei, G., Kloepper, J.W. and Tuzun, S. *Phytopathol* 1991, 81:1508-1512.

118. Zehnder, W.G., Murphy, F.J., Sikora, J.E. et al. *Eur J Plant Pathol*. 2001, 107:39-50.

119. Hyakumachi, M. 1997. *In:* Plant Growth-Promoting Rhizobacteria - Present Status and Future Prospects. Ogoshi, A., Kobayashi, K., Homma, Y. et al. (eds), Sapporo, Fac. Agric. Hokkaido. Univ., 1997, pp. 164–169.

120. Liu, L., Kloepper, J.W. and Tuzun, S. *Phytopathol* 1995, 85:843-847.

121. Bakker, P.A.H.M., Pieterse, C.M.J. and van Loon, L.C. *Phytopathol* 2007, 97:239-243.

122. Gomez-Gomez, L. and Boller, T. *Mol Cell* 2000, 5:1-20.

123. Zipfel, C., Robatzek, S., Navarro, L. et al. *Nature* 2004, 428:764-767.

124. Meziane, H., Van der Sluis, I., Van Loon, L.C. et al. *Mol Plant Pathol* 2005, 6:177-185.

125. Graham, T.L., Sequeira, L. and Huang, T.S.R. *Appl Environ Microbiol* 1977, 34:424-432.

126. Minardi, P., Fede, A. and Mazzucchi, U. *J Phytopathol* 1989, 127:211-220.

127. Van Peer, R. and Schippers, B. *Neth J Plant Pathol* 1992, 98:129-139.

128. Leeman, M., Van Pelt, J.A., Den Ouden, F.M. et al. *Phytopathol* 1995, 85:1021-1027.

129. Van Wees, S.C.M., Pieterse, C.M.J., Trijssenaar, A. et al. *Mol Plant-Microbe Interact* 1997, 10:716-724.

130. Whipps, J.M. *J Exp Bot* 2001, 52:487-511.

131. Compant, S., Duffy, B., Nowak, J. et al. *Microbiol* 2005, 4951-4959.

132. Loper, J.E. and Henkels, M.D. *Appl Environ Microbiol* 1999, 65:5357-5363.

133. O'Sullivan, D.J. and O'Gara, F. *Microbiol Rev* 1992, 56:662-676.

134. Castignetti, D. and Smarelli, J. *FEBS Lett* 1986, 209:147-151.

135. Lodewyckx, C., Vangronsveld, J., Porteous, F. et al. *Crit Rev Plant Sci* 2002, 21:583-606.

136. Wang, Y., Brown, H.N., Crowley, D.E. et al. *Plant Cell Environ* 1993, 16:579-585.

137. Cornelis, P. and Matthijs, S. *Environ Microbiol* 2002, 4:787-798.

138. Ravel, J. and Cornelis, P. *Trends Microbiol* 2003, 11:195-200.

139. Saleh, S.S. and Glick, B.R. *Can J Microbiol* 2001, 47:698-705.

140. Kojic, M., Degrassi, G. and Venturi, V. *Biochim Biophys Acta* 1999, 1489:413-420.

141. Ovadis, M., Liu, X., Gavriel, S. et al. *J. Bacteriol* 2004, 186:4986-4993.

142. Duffy, B.K. and Défago, G. *Appl Environ Microbiol* 1999, 65:2429-2438.

143. Pozo, M.J., Cordier, C., Dumas-Gaudot, E. et al. *J Exp Bot* 2002, 53:525-534.

144. Pozo, M.J. and Azcon-Aguilar, C. *Curr Opinion Plant Biol* 2007, 10:393-398.

145. Raju, P.S., Clark, R.B., Ellis, J.R. et al. *Plant Soil* 1990, 121:165-170.

146. Marschner, H. and Dell, B. *Plant Soil* 1994, 159:89-102.

147. Trotta, A., Varese, G.C., Gnavi, E. et al. *Plant Soil* 1996, 185:199-209.

148. Matsubara, Y., Kayukawa, Y., Yano, M. et al. *J Jpn Soc Hortic Sci* 2000, 69:552-556.

149. Matsubara, Y., Ohba, N. and Fukui, H. *JJpn Soc Hortic Sci* 2001, 70:202-206.

150. Subramanian, K.S. and Charest, C. *Mycorrhiza* 1995, 5:273-278.

151. El-Tohamy, W., Schnitzler, W.H. and El-Behairy, U. *J Appl Bot* 1999, 73:178-183.

152. Ho, I. *Northwest Sci* 1987, 61:148-159.
153. Shaul, O., Galili, S., Volpin, H. et al. *Mol Plant Microbe Interact* 1999, 12:1000-1007.
154. Whipps, J.M. *Can J Bot* 2004, 82:1198-1227.
155. Gernns, H., von Alten, H. and Poehling, H.M. *Mycorrhiza* 2001, 11:237-243.
156. Lingua, G., D'Agostino, G. and Massa, N. *Mycorrhiza* 2002, 12:191-198.
157. Fritz, M., Jakobsen, I., Lyngkjaer, M.F. et al. *Mycorrhiza* 2006, 16:413-419.
158. Liu, J., Maldonado-Mendoza, I., Lopez-Meyer, M. et al. *Plant J* 2007, 50:529-544.
159. Shaul, O., Galili, S., Volpin, H. et al. *Mol Plant Microbe Interact* 1999, 12:1000-1007.
160. Cordier, C., Pozo, M.J., Barea, J.M. et al. *Mol Plant Microbe Interact* 1998, 11:1017-1028.
161. Hahn, M.G., Bucheli, P., Cervone, F. et al. *In:* Plant-Microbe Interactions: Molecular and Genetic Perspectives. Kosuge T, Nester EW (eds.), Macmillan, New York, 1989, pp. 131-181.
162. Collinge, D.B., Gregersen, P.L. and Thordal-Christensen, H. *In:* Mechanisms of Plant Growth and Improved Productivity: Modern Approaches and Perspectives. Basra AS (ed.), Marcel Dekker, New York, 1994, pp. 391-433.
163. Kuæ, J. *In:* Innovative Approaches to Plant Disease Control. Chet I. (ed.), J. Wiley, Sons Inc, New York, 1987, pp. 255-274.
164. Gottstein, H.D. and Kuæ, J.A. *Phytopathol* 1989, 79:176-179.
165. Reuveni, M., Agapov, V. and Reuveni, R. *Plant Pathol* 1995, 44:31-39.
166. Oostendrop, M., Kunz, W., Dietrich, B. et al. *Eur J Plant Pathol* 2001, 197:19-28.
167. Mucharromah, E. and Kuæ, J. *Crop Prot* 1991, 10:265-270.
168. Sticher, L., Mauch-Mani, B. and Metraux, J.P. *Annu Rev Plant Pathol* 1997, 35:235-270.
169. Cohen, Y., Gisi, U. and Moesinger, E. *Physiol Mol Plant Pathol* 1991, 38:255-263.
170. Gamil Nagwa, A.M. *Ann Agric Sci Moshtohor* 1995, 33:681-691.
171. Mosa, A.A. *Ann Agri Sci Ain Shams Univ Cairo* 1997, 42:241-255.
172. Mosa, A.A. *Arab Univ J Agric Res* 2002, 10:1043-1057.
173. Mosa, A.A. *Ann Agric Sci Ain Shams Univ Cairo* 2002, 47:993-1008.
174. Matsumoto, I., Ohguchi, T., Inoue, M. et al. *Ann Phytopathol Soc Japan* 1978, 44:22-27.
175. Mills, P.R. and Wood, S.K.R. *Phytopathol Z* 1983, 111:209-216.
176. Dean, R.A. and Kuc, J. *In:* Fungal infection of plants. Page, F. and Qyers, P.G. (eds), Cambridge University Press, Cambridge, 1987, pp. 383-410.
177. Kessmann, H., Staub, T., Hofmann, C. et al. *Annu Rev Plant Pathol* 1994, 32:439-459.
178. Aly, M.M. *Egypt Soc Appl Microbiol* 1989, 316-328.
179. Abou El-Hawa, M. *Egypt J Microbiol* 1998, 33:147-154.
180. Kunz, W., Schurter, R. and Maetzke, T. *Pestic Sci* 1997, 50:275-282.
181. Doubrava, N., Dean, R. and Kuæ, J. *Physiol Mol Plant Pathol* 1988, 33:69-79.
182. Avdiushko, S.A., Xs, Ye and Kúc, J. *Physiol Mol Plant Pathol* 1993, 42:441-454.
183. Chen, Z.P., Kloek, A., Boch, J. et al. *Mol Plant Microbe Interact* 2000, 13:1312-1321.
184. Wicks, T.J., Magarey, P.A., Wachtel, M.F. et al. *Plant Dis* 1991, 75:40-43.

7

Influence of VAM in Bioremediation of Environmental Pollutants

V. Davamani, A.C. Lourduraj[1] and M. Velmurugan[2]

Agro-Climate Research Centre, Tamil Nadu Agricultural University
Coimbatore 641003, Tamil Nadu, India

[1]Water Technology Centre, Tamil Nadu Agricultural University
Coimbatore 641003, Tamil Nadu, India

[2]Horticultural College and Research Institute, Tamil Nadu Agricultural
University, Coimbatore 641003, Tamil Nadu, India

Introduction

Environmental degradation due to the disposal of industrial and urban wastes generated by human activities has become a major environmental concern. Controlled and uncontrolled disposal of wastes to agricultural soils are responsible for the migration of contaminants into non-contaminated sites [1]. Soil contamination by heavy metals may pose a threat to human health, if the metals enter the food chain [2]. Soil remediation is therefore needed to eliminate risk to humans from these toxic metals [3]. Biosphere pollution by heavy metals and nucleotides was accelerated dramatically during the last few decades due to mining, smelting, manufacturing, treatment of agricultural soils with agro-chemicals and soil sludge, etc. Problems associated with the contamination of soil and water such as animal welfare, health, fatalities and disruptions of natural ecosystems are well documented [4]. Heavy metals such as Pb, Cr, As, Cu, Cd, and Hg, being added to our soils through industrial, agricultural and domestic effluents, persist in soils and can either be adsorbed in soil particles or leached into ground water. Human exposure to these metals through ingestion of contaminated food or uptake of drinking water can lead to their accumulation in humans, plants and animals. Lead, copper, zinc and cadmium are also found naturally in soils and they can cause significant damage to environment and human health as a result of their mobility and solubilities. They can occur in soil and water in several forms and their speciation in soils are determined by sequential extraction using specific extractants, which solubilize different phases of metals [5]. The physical and chemical characteristics of soil determine the speciation and mobility of heavy metals [6].

Heavy metals are highly persistent in soil over long period, with consequent problems for the human population. Besides, numerous heavy metals were found to be toxic to plant growth and production [7]. Soil microorganisms are important in the recovery of the potentially toxic environments and might be used in agriculture to remediate the polluted soil and improve nutrient availability to plants [8]. Arbuscular mycorrhizal fungi (AMF) are known to improve plant growth on nutrient-poor soils and enhance their uptake of P, Cu, Ni, Pb and Zn [9, 10]. Number of reports are available for extraction of metals from the soil [9]. Unfortunately, most of the accumulative plants used belong to the family Brassicaceae which rarely form AM symbiosis, besides, these plants produce little biomass. Plants with higher biomass production such as trees are of more interest in soil phytoremediation [11]. Besides, many tree species growing on metal-contaminated soils possess mycorrhizae [12], indicating that these organisms have evolved a tolerance to heavy metals and can play an important role in the phytoremediation of contaminated soils [9, 13]. *Eucalyptus rostrata* trees have a high capacity to grow in poor or marginal soils [14]. This species is able to develop mycorrhizal symbiosis and has a great tolerance to heavy metals [15]. AM fungi were able to increase *Phaseolus vulgaris* growth in heavy metal contaminated soils and found to reduce their harmful effect on plants [16]. Under field conditions, different plant species live together and hyphae of AM fungi interconnect the root systems of adjacent plants and can mediate nutrient transfer between plants [17].

In many terrestrial ecosystems, arbuscular mycorrhizae (AM) and dark septate fungi (DSF) colonisation also vary according to seasonal patterns [18] and the facilitating role of mycorrhizae in tolerance and metal root uptake has been hypothesized by several workers. The prospect of AM fungi existing in heavy metal-contaminated soils as well as used to reclamation of sites contaminated by various polluting agents has important implications for phytoremediation. Since heavy metal uptake and tolerance depend on both plant and soil factors, including soil microbes, interactions between plant root and their symbionts such as AM fungi can play an important role in successful survival and growth of plants in contaminated soils. Mycorrhizal associations increase the absorptive surface area of the plant due to extrametrical fungal hyphae exploring rhizospheres beyond the root-hair zone, which in turn enhances water and mineral uptake. AM fungi can further serve as a filtration barrier against transfer of heavy metals to plant shoots. The protection and enhanced capability of uptake of minerals result in greater biomass production, a prerequisite for successful remediation. Indigenous AM isolates existing naturally in heavy metal-polluted soils are more tolerant than isolates from non-polluted soils, and are reported to efficiently colonize plant roots in heavy metal-stressed environments. Thus, it is important to screen indigenous and heavy metal-tolerant isolates in order to guarantee the effectiveness of AM symbiosis in restoration of contaminated soils. It is further suggested that the potential of phytoremediation of contaminated soil can be enhanced by inoculating hyper-accumulator plants with mycorrhizal fungi most appropriate for the contaminated site.

Mycorrhizal fungi can contribute to the improving of microbial population in degraded soil, plant growth, immobilization of nutrients, soil aggregation, binding of pollutants etc. However, AM fungal isolates adapted to local soil conditions can stimulate plant growth better than non-indigenous isolates. Indigenous AM fungal ecotypes result from long-term adaptation to soils with extreme properties [19]. Therefore, isolation of indigenous stress-adapted AM fungi can be a potential biotechnological tool for inoculation of plants in disturbed ecosystems [20]. About 95% of the world's plant species belong to characteristically mycorrhizal families [21] and potentially benefit from AM fungus-mediated mineral nutrition [22] due to the fundamental role played by these glomalean fungi in biogeochemical element cycling [23]. AM symbiosis occurs in almost all habitats and climates [24], including disturbed soils [25, 26] and those derived from mine activities [27, 28]. The fungi can accelerate the revegetation of severely degraded lands such as coal mines or waste sites containing high levels of heavy metals [29, 30]. AM fungi form an integral component of successfully revegetated flue-gas-desulphurization sludge ponds.

Needs of Mycorrhizal Fungi for Removal of Pollutants

Plants which appear spontaneously in polluted places are frequently devoid of mycorrhizal symbiosis and are mostly characterized by poorly developed root and shoot biomass when heavy metals are present [31]. The lack of mycorrhiza can hamper the revegetation of the metal-contaminated mine spoil or other degraded sites. The introduction of an AM fungal inoculum into these areas could be a strategy for enhancing the establishment of mycorrhizal herbaceous species. AM fungal isolates differ in their effect on heavy metal uptake by plants [32]. Some reports indicate higher concentrations of heavy metals in plants due to AM [33], whereas others have found a reduced plant concentration; for example, Zn and Cu in mycorrhizal plants [34]. Thus, selection of appropriate isolates could be of importance for a given phytoremediation strategy. AM fungal species can be isolated from areas which are either naturally enriched by heavy metals or are old mine/industry waste sites in origin. In this context, AM fungi constitute an important functional component of the soil-plant system that is critical for sustainable productivity in degraded soils. AM fungi are of importance as they play a vital role in metal tolerance [35] and accumulation [36, 37]. External mycelium of AM fungi provides a wider exploration of soil volumes by spreading beyond the root exploration zone [38, 39], thus providing access to greater volume of heavy metals present in the rhizosphere. A greater volume of metals is also stored in the mycorrhizal structures in the root and in spores. For example, concentrations of over 1200 mg kg^{-1} of Zn have been reported in fungal tissues of *Glomus mosseae* and over 600 mg kg^{-1} in *G. versiforme* [40]. Another important feature of this symbiosis is that AM fungi can increase plant establishment and growth despite high levels of soil heavy metals, due to better nutrition [41, 42], water availability [43] and soil aggregation properties [44] associated with this symbiosis. AM fungus is significant in the ecological improvement of rhizosphere [45, 46].

Role of Mycorrhizal Fungi in Rhizosphere

Glomales comprise one of the oldest groups of fungi, older than land plants. The first land plants, Bryophytes, appeared in the Mid Silurian era. The oldest fossil evidence of Bryophyte-like land plants (approximately 100 years ago) had AM-like infections even before roots evolved [47]. The first land plants were most likely evolved from algae but no fossil records are available for the rootless freshwater Charophycean algae, which are the probable ancestors of land plants, to show if they were mycorrhizal. Mosses, liverworts, and hornworts often contain structures like hyphae, vesicles, and arbuscules, characteristics of AM fungi [48]. *Sphenophytes*, *Lycopodophytes* and *Pteridophytes* are among the first land plants with roots which originated in the mid Devonian era, and AM associations are reported in these plants [49]. Both living and Triassic fossil Cycades had AMF in their roots. AM associations are ubiquitous in the living angiosperms, which probably arose in the early Cretaceous era [50]. It is hypothesized that AM fungi were instrumental in the colonization of land by ancient plants [51]. This hypothesis is supported by observation that AM can now be found worldwide in the angiosperms, gymnosperms as well as ferns, suggesting that the nature of the association is ancestral. Furthermore, the origin of AM fungi coincides with that of vascular plants suggesting the nature of the association and supporting the hypothesis that AMF were instrumental in the colonization of land by ancient plants.

It is now established that universal and ubiquitous symbiotic arbuscular mycorrhizal (AM) fungi, belonging to Glomales, form symbiotic relationships with roots of 80-90% land plants in natural and agricultural ecosystems [49], including halophytes, hydrophytes and xerophytes [52], and are known to benefit plant nutrition, growth and survival, due to their greater exploitation of soil for nutrients [53]. These associations represent a key factor in the below-ground networks which influence diversity and plant community structure [54], but we know very little about the enormous AM fungal diversity in soils and their properties and behaviour in the soil [55]. The degree of benefit to each partner in any AMF-plant host interaction depends not only on the particular plant and AMF species involved but also on the rhizobacteria and soil abiotic factors.

Mechanism of Heavy Metal Uptake by Mycorrhizal Fungi

Heavy metals in soil are associated with a number of soil components which determine their behaviour in the soil and influence their bioavailability [56]. The cell wall components such as free amino, hydroxyl, carboxyl and other groups of soil fungi can bind with potentially toxic elements such as Cu, Pb, Cd, etc. [57]. Many filamentous fungi can sorb these trace elements and are used in their commercial biosorbants [58]. The proteins in the cell walls of AM fungi appear to have similar ability to sorb potentially toxic elements by sequestering them. There is evidence that AMF can withstand potentially toxic elements. Gonzalez-Chavez et al. [59] showed that glomalin produced on hyphae of AMF can sequester them. AMF play a significant ecological role in the phytostabilization of potentially toxic trace element polluted soils by sequestration and in turn, help mycorrhizal plants survive in polluted soils. One of these components is glomalin, a glycoprotein produced by the hyphae of AMF fungi [60], which is released into soil from AMF hyphae [61]. These authors, using an *in vitro* system of Ri T-DNA transformed roots infected with *Glomus intraradices*, an AMF, showed that glomalin is tightly bound in AMF hyphal and spore walls. Small amounts (<20%) of glomalin were found to be adhered to soil via release into liquid medium from hyphae and not through passive secretion and their function is physiological in the course of the life of the organism. It has been hypothesized that glomalin has a role in the immobilization (filtering) of heavy metals at the soil-hypha interface, i.e. before entry into fungal-plant system. The extra-radical mycelium of AMF, in addition to its crucial role in enhancing nutrition of host plant, also plays a role in soil particle aggregation and soil stability [60, 62]. There has been few analytical studies of AM in polluted soils. While some workers observed that the external mycelium of AMF was the main site for trace element localization [63], others reported selective exclusion of toxic and non-toxic elements by adsorption onto chitinous cell walls or onto extra-cellular glycoprotein, glomalin [60], or intra-cellular precipitation. All these mechanisms have implications in reducing a plant's exposure to potentially toxic elements, i.e. mycorrhizoremediation technology. Gonzalez-Chavez et al. [64] studied the form and localization of Cu accumulation in the extra-radical mycelium of three AM fungi isolated from the same polluted soil contaminated with Cu and As. The authors reported differential capacity of AMF to sorbs and accumulate Cu as determined by TEM and SEM. However, the nature of accumulation and mechanisms involved require further studies in order to better understand the participation of AMF in plant tolerance and its ecological significance in polluted soils.

The AMF can be screened for their ability to produce maximum levels of extra-radical mycelium in polluted soils [65], and to utilize adapted AMF to help accumulate heavy metal both within the plant roots (phytoaccumulation) and the extra metrical fungal mycelium. Most of the information available in literature concerning fairly recently discovered soil compound, glomalin, an iron-containing glycol-soil-proteinaceous substance produced by AMF, is in relation to its role in soil aggregation [66]. Glomalin makes a large contribution to active soil organic C pools. It is quite recalcitrant and we know very little about its chemical structure. Glomalin plays a vital part in sorption and sequestration of potentially toxic elements, reducing their bioavailability. It has been suggested that this sequestration could be important for heavy metal biostabilization in heavy metal polluted soils [67]. Glomalin attaches to soil and helps stabilize aggregates. Mycorrhizae may be of great significance for the survival of plants at HM polluted sites. Literature data on glomalin content in HM polluted soils are extremely scarce and there is a need to gather data on this aspect of mycorrhizoremediation, i.e., use of mycorhizal plants in the phytoremediation of HM contaminated soils.

Several biological and physical mechanisms have been proposed to explain metal tolerance of AM fungi and AM fungal contribution to metal tolerance of host plants. Immobilization of metals in the fungal biomass is one such mechanism involved [68]. Reduced transfer, as indicated by enhanced root/shoot Cd ratios in AM plants, has been suggested as a barrier in metal transport [69, 70]. This may occur due to intracellular precipitation of heavy metal cations with PO_4^- ions. Turnau et al. [71] have demonstrated greater accumulation of Cd, Ti and Ba in fungal structures than in the host plant cells. Uptake into hyphae may be influenced by absorption on hyphal walls as chitin has an important metal-binding capacity [72]. Thus, AM fungal metal tolerance includes adsorption onto plant or fungal cell walls present on and in plant tissues, or onto or into extraradical mycelium in soil [73], chelation by such compounds as siderophores and metallothionins released by fungi or other rhizosphere microbes, and sequestration by plant-derived compounds like phytochelatins or phytates. Other possible metaltolerance mechanisms include dilution by increased root or shoot growth, exclusion by precipitation onto polyphosphate granules, and compartmentalization into plastids or other membrane-rich organelles. Indirect mechanisms include the effect of AM fungi on rhizosphere characteristics such as changes in pH [74], microbial communities [75] and root-exudation patterns [76].

Role of Mycorrhizal Fungi in Bioremediation

VAM are a useful and important microorganism in the hot and arid environment for the bioremediation of polluted soils. The effect of inoculation of vesicular arbuscular mycorrhiza (VAM) on the height of selected ornamental plants in field condition was evaluated and found that the efficiency of the inoculation is significantly higher in the initial stages, which means the application can be easily and successfully done at the nursery stages. In both bioremediated and agricultural soils, VAM inoculated plants performed better than the uninoculated plants. The possible use of mycorrhizal fungi as bio indicators of polluted sites must be supported by consistent field studies. Friedel et al. [77] studied the effect of waste water irrigation on soil organic matter, soil microbial biomass and microbial activities. In the case of AM fungi, pollution impact will occur also on physiological level (symbiotic efficiency). Since pollutants are added constantly and at low rates, allowing an adaptation of the organism to pollution and therefore not reflecting only on spore abundance.

Role of Mycorrhizal Fungi in Aromatic Hydrocarbon Degradation

An important class of aromatic hydrocarbons is represented by benzene, toluene, ethylbenzene, meta-, para- and ortho-xylene (BTEXs). They constitute a significant percentage (up to 18%) of many petroleum derivatives [78]. The spilling of gasoline and diesel fuel from underground leaking oil tanks, in oil distribution and storage stations, is one of the most common sources of BTEX pollution in soils [79]. The risk for human health is related to their acute and long-term toxic effects, including genotoxicity and carcinogenicity. Biodegradation of BTEX aromatic hydrocarbons by fungi is well documented for non-symbiotic, saprotrophic species [80]. Positive interactions between AM and beneficial pseudomonads have been shown, and bacteria phylogenetically close to pseudomonads are known to be obligate endosymbionts of *Gigaspora margarita* [81].

Arbuscular mycorrhizal (AM) fungi have been widely studied as a potential tool in the reclamation of sites contaminated by various polluting agents [82]. The only available information about organic pollutants deals with the effect of AM on the fate of polycyclic aromatic hydrocarbons (PAH) in polluted soils [83]. Root colonization by *G. rosea* and *G. mosseae* was generally maintained high in the presence of BTEX, whilst much lower colonization was observed for *G. margarita* even in the absence of pollutants. The effect of BTEX exposure on root colonization varied with the type of pollutant and the AM fungus. Decreases in root colonization and arbuscule formation by *G. mosseae* were observed in the presence of xylene and, particularly, benzene, whilst *G. rosea* was more sensitive to toluene and ethylbenzene. Such detrimental influence of organics on AM colonization has previously been documented [84].

Mycorrhizal Fungi Degradation of Organic Contaminants

Most previous studies have found that AM fungi have positive effects on the dissipation of organic contaminants such as atrazine [85], PAHs [86], DDT [87] and weathered p,p-DDE in soils [88] and a depression in PAH dissipation in the presence of ectomycorrhizas has also been reported. AM fungi may therefore play a critical role in the degradation of organic contaminants in soils.

The mechanisms involved in interactions between AM fungi and organic contaminants in soil remain unclear. It is reasonable to expect that soil microbial activity enhanced and soil microbial communities modified by AM fungi play a key role in the degradation of organic contaminants. Once arbuscular mycorrhizal association has developed, AM hyphae influence the surrounding soil which has been termed the mycorrhizosphere, resulting in the development of distinct microbial communities in the rhizosphere and bulk soil [89]. Phospholipid fatty acid (PLFA) analysis has revealed an important qualitative difference in microbial community structure in mycorrhizosphere soil as affected by AM fungi in PAH-spiked soil [90]. Furthermore, the AM fungal hyphosphere, the zone of soil affected by the extraradical hyphae may support a distinct microbial community within the mycorrhizosphere and exert effects on degradation of organic compounds in soil.

Studies have indicated that AM fungi can increase the activities of soil enzymes such as phosphatase and dehydrogenase. Dehydrogenase, a soil oxidoreductase, is an intracellular enzyme catalyzing oxidoreduction reactions of organic compounds. Several studies have demonstrated that the dehydrogenase enzyme activity of microorganisms is one of the most sensitive parameters available for toxicity evaluation and alkaline phosphatase is involved in the process of phosphate acquisition in mycorrhizal plants. Both of these enzymes are considered to play key metabolic roles in mycorrhizal function [91]. However, there have been few reports that deal with the effects of AM inoculation on soil enzymes and microbial community structure in soils containing organic contaminants. The studies reported that atrazine dissipation in soil was enhanced by AM inoculation [92]. The major reason for increased dissipation of atrazine in soil might contribute to: (1) specific effects of mycorrhizal roots and extraradical mycelium on atrazine dissipation; and/or (2) modification of the effects on soil enzyme activities and microbial community structure resulting from mycorrhizal inoculation.

Role of Mycorrhizal Fungi in Phytoremediation

Phytoremediation is cost effective and a good alternative to the conventional chemical and physical methods of treating contaminated soils [93]. Phytoremediation has been successfully used to remove heavy metals from soils. Mycorrhizae have also been reported in plants growing on heavy metal-contaminated sites indicating that these fungi have evolved a HM-tolerance and that they may play a role in the phytoremediation of the site. Joner and Leyval [94] found that cadmium-tolerant *Glomus mosseae* isolates AMF were responsible for uptake, transport and immobilization of cadmium. Copper (Cu) was absorbed and accumulated in the extraradical mycelium of three AMF isolates, as observed in a study with *Glomus* spp. Mycorrhizae were found to ameliorate the toxicity of trace metals in polluted soils growing in soybean and lentil plants. As mycorrhizae may enhance the ability of the plant to cope with water stress situations associated to nutrient deficiency and drought, mycorrhizal inoculation with suitable fungi has been proposed as a promising tool for improving restoration success in semi-arid degraded areas.

Tolerant Capacity of Mycorrhizal Fungi with Heavymetals

The earlier studies reported the abundance of AM fungi in long-term sewage-sludge field trials, where contaminated sludge was applied over a period of 18 years [95]. *Glomus claroideum*, an ecotype, isolated from plots receiving 300 $m^3 ha^{-1} yr^{-1}$ of contaminated sludge was potentially adapted to metal tolerance in contaminated soils [96]. This isolate produced a significantly higher root colonization level (20% in *Allium porrum* and 15% in *Sorghum bicolor*) in the heavy metal-contaminated soil than produced by other isolates (isolated from non-contaminated soil). Furthermore, *G. claroideum* and *Glomus constrictum*, were the most effective fungus in improving the growth of *A. porrum* and *S. bicolor* in heavy metal-contaminated soil. Spores from naturally polluted soils (collected from contaminated sites close to a Zn smelter and a site overlying with alum shale bedrock) have been shown to germinate better in heavy metal-polluted soil, compared to spores from non-polluted soils [97] [Table 1]. Thus, indigenous AM fungi showed tolerance to heavy metals as they germinated in soils containing 6060 $mgkg^{-1}$ Pb, 24,410 $mgkg^{-1}$ Zn and 1630 $mgkg^{-1}$ Cu. Weissenhorn et al. [98] reported the abundance of AM fungi (100 spores per 50 g soil) in two agricultural soils close to a Pb-Zn smelter. Spores belonging to the *G. mosse*e group were isolated from two heavy metal-polluted soils in

Table 1. Effect of mycorrhizal fungi in phytoremediation

Plants species	Fungal species	Heavy metals	Reference
A. capillaris, Zea mays, Lygeum spartum	*Glomus intraradices, G. mosseae, Glomus macrocarpum*	Pb	[114]
Berkheya coddii, A. porrum, Sorghum bicolor	*Gigaspora* sp., *Glomus tenue, G. caledonium*	Ni	[115]
Trifolium repens, Hordeum vulgare, Trifolium subterraneum, Allium porrum	*Glomus mosseae, Glomus* sp., *Gigaspora* sp., *G. mosseae*	Cd	[33, 98, 116, 117]
Viola calaminaria, Trifolium repens, Festuca rubra, Calamagrostis sp.	*Glomus* sp., *Glomus constrictum, Glomus ambisporum, Glomus fasiculatum*	Zn	[37, 118, 119]

France. The two cultures isolated from Cd-polluted soils were found to be tolerant to Cd concentrations of approximately 50–70 µg l^{-1} and 200-500 µg l^{-1} respectively. Highly functional AM symbiosis was reported from southern Poland, in plants colonizing calamine spoil mounds rich in Cd, Pb and Zn. Mycorrhizal colonization was much higher in *Plantago lanceolata* (up to 90%) compared to *Biscutella laevigata* (up to 40%). Besides vesicles and coils, arbuscules were also observed.

Several heavy metal-tolerant AM fungi have been isolated from polluted soils, which can be useful for reclamation of such degraded soils as they are found to be associated with a large number of plant species in heavy metal-polluted soil. Gildon and Tinker [99] isolated a mycorrhizal strain which tolerated 100 mg kg^{-1} of Zn in the soil. Considerable amount of AM fungal colonization was also reported in an extremely polluted metal mining area with HCl-extractable Cd soil concentration of more than 300 mg kg^{-1} [100]. Similarly, Weissenhorn et al. [98] isolated mycorrhizal fungi from two heavy metal-polluted soils, which were found to be more resistant to Cd than a reference strain. Sambandan et al. [101] reported 15 AM fungal species from heavy metal-contaminated soils from India. Of the 15 AM species isolated, *Glomus geosporum* was encountered in all the sites studied. The percentage colonization ranged from 22 to 71% and spore count was as high as 622 per 100 g soil. Turnau et al. [102] analysed the community of AM fungi in roots of *Fragaria vesca* growing in Zn-contaminated soil. Seventy per cent of the root samples containing positively stained fungal hyphae were found to be colonized by *G. mosseae*. Another unique AM fungal species, *Scutellospora dipurpurascens* has been reported by Griffioen et al. [103] from the rhizosphere of *Agrostis capillaris* growing in contaminated surroundings of a zinc refinery in the Netherlands. This indicates that these fungi have evolved Zn and Cd tolerance and that they might play an important role in conferring Zn and Cd tolerance in plants.

Different AM fungi have been shown to differ in their susceptibility and tolerance to heavy metals. del Val et al. [35] reported six AM fungal ecotypes with consistent differences with regard to their tolerance to the presence of heavy metals. AM fungal ecotypes ranged from very sensitive to the presence of metals to relatively tolerant to high rates of heavy metals in soil. *Glomus* sp. (isolated from non-polluted soil) was shown to be the most sensitive fungus, while one strain of *G. claroideum* (isolated from the most contaminated soil) was the most tolerant. The effectiveness of the different AM fungal isolates in improving plant growth also depends on the level of heavy metals in soil. Furthermore, AM fungi from different soils may differ in their metal susceptibility and both metal-specific and nonspecific tolerance mechanisms may be selected in metal-polluted soil [104].

Joner and Leyval [33] reported that a large proportion of increased Cd content of mycorrhizal plants was sequestered in the roots. The mycorrhizal status of the plants did not influence shoot concentration of Cd, but concentration in roots was increased in mycorrhizal plants. Cd-root:shoot ratio was found to be 3.15 in mycorrhizal plants compared to 1.66 in non-mycorrhizal plants. They concluded that extraradical hyphae of AM fungus can transport Cd from soil to plant, but transfer from fungus to plant is restricted due to fungal immobilization. In other experiments [105], subterranean clover was grown in pots containing a soil supplemented with Cd and Zn salts and inoculated with AM fungal spores extracted from the rhizosphere of *V. calaminaria* and spores of Cd-tolerant *G. mosseae* [Table 1]. The AM fungal inoculum from the metal-tolerant plant (*V. calaminaria*) was efficient in sequestering metals in the roots. Clover roots inoculated with AM fungi from the rhizosphere of this plant contained eightfold higher Cd and threefold higher Zn concentrations compared to non-inoculated plants, without any significant difference in plant biomass and concentration of metals in shoots. Hildebrandt et al. [106] reported that *Glomus* Br1 isolated from the roots of *V. calaminaria* improved maize growth in a polluted soil and reduced root and shoot heavy-metal concentration in comparison to a common *Glomus intraradices* isolate or non-colonized controls. Cd-tolerant *G. mosseae* from metallophyte plant enhanced growth of maize, alfalfa, barley, etc. in heavy metal-rich soils. Thus, mycorrhizal fungi adapted to elevated soil metal concentration can significantly improve the growth and plant P nutrition under metal stress. By maintaining a higher shoot P/Zn concentration ratio mycorrhizal plants are able to alleviate the negative effects of Zn [107].

The higher heavy metal concentration in mycorrhizal plants could be explained by the fact that AM infection increased plant uptake of metals by mechanisms such as enlargement of the absorbing area, volume of accessible soil and efficient hyphal translocation. The AM fungus, *G. deserticola* and *Glomus macrocarpum* are suitable AM fungus to remove high quantities of Pb from contaminated soils and a correlation between the level of AM colonization and the quantity of Pb absorbed by *Eucalyptus* was found [Table 1]. Arriagada et al. [108] found a positive effect of *G. deserticola* on plant growth and a higher tolerance of the mycorrhizal plants to Pb toxicity. Higher Pb accumulation in the roots than in the shoots of *Eucalyptus* has been observed. Rabie [109] suggested that the symbioses provided the host with nutrients such as phosphorus which may be involved in plant Pb detoxification by means of molecules of phytates that can neutralize excess metals, or P can provide metabolic energy indirectly as ATP for possible compartmentalization within the cell vacuoles by means of molecules such as metallothioneins and phytochelatins.

Most fertilizers, especially the inorganic ones, have heavy metals like Hg, Cd or Cr as impurities. Therefore, addition of these fertilizers to enhance metal solubility, mobility and bioavailability in phytoremediation [110], inadvertently is adding to the quantities of these heavy metals in soil. The more the quantities applied for enhancement, the more their accumulation in the soil ecosystem.

Good enough, among the rhizosphere microorganisms involved in plant interactions with soil are the arbuscular mycorrhiza fungi [111]. Therefore, phytoremediation, enhanced by arbuscular mycorrhiza (AM) fungi offers a natural, efficient and cost effective means of soil remediation. Harrison [112] observed that AM fungi which are natural constituents of the soil ecosystem have ability to interact with plants' roots thereby enlarging the soil volume for better nutrient uptake. Plants that are capable of forming this association with AM fungi have been shown to accumulate considerable amount of trace metals. Awotoye [113] also observed that these fungi help to improve plant growth in soils with low fertility level while at the same time enhancing the uptake of phosphorus.

Conclusion

Mycorrhizal fungi are present widely in the ecosystem, which provide mutual symbiosis of the almost all higher plants and play a major role in uptake of nutrient as well as different type of pollutants. Mycorrhizal fungi improve the nodulation and stimulate the PGPR growth. The application of mycorrhizal fungi protects host plant from the potential toxicity caused by heavy metal. These fungi also help to grow and rehabilitate pollutant contaminated site. The fungi associated with host plant help to improve the resistance from pollutant and heavy metals. A number of studies have shown that same plant species can respond differently to the presence of different mycorrhizal fungal species with various kinds of pollutants. Based on this, it would be interesting to evaluate further combination of mycorrhizal fungi and host plant for their influence on persistence pollutant in contaminated site. Despite the importance of the role that AM play in soil-microbe-plant interactions, relatively few studies have focused on their potential in phytoremediation efforts. This is first due to the fact that earlier phytoremediation studies focused on the predominantly non-mycorrhizal plant families such as Brassicaceae and Caryophyllaceae, and second AM have not been considered by earlier workers as important component of phytoremediation practices. It is possible to improve the phytoremediation capabilities of plants by inoculating them with appropriate AM fungi.

In future, efforts need to be made mainly on the following aspects for improving the alleviation of metals from polluting soils: (1) Combined selected plant species with specific mycorrhizal fungi isolates adopted to high concentration of heavy metals; (2) Need to develop new methods and optimize the conditions to grow plant species with highly affected soil with metals; (3) Study the metal speciation with respect to particular host plant and mycorrhizal fungi; and (4) Develop genetically improved plant species for heavy loading of metals in their parts.

References

1. Ghosh, M. and Singh, S.P. *Applied Ecol Environ Res* 1995, 3(1):1-8.
2. Berti, W.R. and Jacobs, L.W. *J Environ Qual* 1996, 25:1025-1032.
3. Lazat, M.M. *J Environ Qual* 2002, 31:109-120.
4. He, Z.L., Yang, X.E. and Stoffella, P.J. *J Trace Elem Med Biol* 2005, 19(2-3): 125-140.
5. Shuman, L.M. *Soil Science* 1985, 140(1):11-22.
6. Kabata-Pendias, A. and Pendias, H. CRC Press, Boca Raton, 1992.
7. Adriano, D.C. Springer, New York, 1986.
8. Shetty, K.G., Hetrick, B.A.D., Figge, D.A.H. et al. *Environ Poll* 1994, 86:181-188.
9. Khan, A.G., Kuek, C., Chaudhry, T.M. et al. *Chemosphere* 2000, 41:197-207.
10. Zhu, Y.G., Christie, P. and Laidlaw, A.S. *Chemosphere* 2001, 42:193-199.
11. Landberg, T. and Greger, M. *Appl Geochem* 1996, 11:175-180.
12. Chaudhry, T.M., Hayes, W.J., Khan, A.G. et al. *Aust J Ecotox* 1998, 4:37-51.
13. Khan, A.G. *Environ Intern* 2001, 26:417-423.
14. Pyatt, F.B. *Ecotox Environ Saf* 2001, 50:60-64.
15. Arriagada, C.A., Herrera, M.A., Garcia-Romera, I. et al. *Symbiosis* 2004, 36:285-299.
16. Heggo, A., Angle, J.S. and Chaney, R.L. *Soil Biol Biochem* 1990, 22:865-869.
17. Bethlenfalvay, G.J., Schreiner, R.P., Mihara, K.L. et al. *Appl Soil Ecol* 1996, 3:205-214.
18. Lingfei, L., Yang, A. and Zhao, Z. *FEMS Microbiol Ecol* 2005, 54:367-373.
19. Sylvia, D.M. and Williams, S.E. *In:* Mycorrhizae in Sustainable Agriculture. Bethlenfalvay, G.J. and Linderman, R.G. (eds), ASA No. 54, Madison, USA, 1992, pp. 101-124.
20. Dodd, J.C. and Thompson, B.D. *Plant Soil* 1994, 159:149-158.
21. Smith, S.E. and Read, D.J. Mycorrhizal Symbiosis. Academic Press, San Diego, USA, 1997.
22. Jeffries, P. and Barea, J.M. *In:* Impact of Arbuscular Mycorrhizas on Sustainable Agriculture and Natural Ecosystems. Gianinazzi, S. and Schuepp, H. (eds), Birkhauser, Basel, 1994, pp. 101-115.
23. Barea, J.M., Azcon-Aguilar, C. and Azcon, R. *In:* Multitrophic Interactions in Terrestrial Systems. Gange, A.C. and Brown, V.K. (eds), Cambridge, 1997, pp. 65-77.
24. Enkhtuya, B., Rydlová, J. and Vosátka, M. *Appl Soil Ecol* 2002, 14:201-211.
25. Estaún, V., Savé, R. and Biel, C. *Appl Soil Ecol* 1997, 6:223-229.
26. Bi, Y.L., Li, X.L., Christie, P. et al. *Chemosphere* 2003, 50:863-869.
27. Bundrett, M.C., Ashwath, N. and Jasper, D.A. *Plant Soil* 1996, 184:173-184.
28. Marx, D.H. *Ohio J Sci* 1975, 75:88-297.
29. Marx, D.H. and Altman, J.D. *In:* *Mycorrhizal Manual*, Springer, Berlin, 1979, pp. 387-399.
30. Wilson, G.W.T., Hetrick, B.A.D. and Schwab, A.P. *J Environ Qual* 1991, 20:777-783.
31. Pawlowska, T.B., Blaszkowski, J. and Rühling A. *Mycorrhiza* 1996, 6:499-505.
32. Leyval, C., Turnau, K. and Haselwandter, K. *Mycorrhiza* 1997, 7:139-153.
33. Joner, E.J. and Leyval, C. *New Phytol* 1997, 135:353-360.

34. Heggo, A., Angle, J.S. and Chaney, R.L. *Soil Biol Biochem* 1990, 22:865-869.
35. del Val, C., Barea, J.M. and Azcòn-Aguilar, C. *Appl Soil Ecol* 1999, 11:261-269.
36. Jamal, A., Ayub, N., Usman, M. et al. *Int J Phytoremed* 2002, 4:205-221.
37. Zhu, Y.G., Christie, P. and Laidlaw, A.S. *Chemosphere* 2001, 42:193-199.
38. Khan, A.G., Kuek, C., Chaudhry, T.M. et al. *Chemosphere* 2000, 41:197-207.
39. Malcova, R., Vosátka, M. and Gryndler, M. *Appl Soil Ecol* 2003, 23:55-67.
40. Chen, B., Christie, P. and Li, L. *Chemosphere* 2001, 42:185-192.
41. Feng, G., Song, Y.C., Li, X.L. et al. *Appl Soil Ecol* 2003, 22:139-148.
42. Taylor, J. and Harrier, L.A. *Appl Soil Ecol* 2001, 18:205-215.
43. Auge, R.M. *Mycorrhiza* 2001, 11:3-42.
44. Kabir, Z. and Koide, R.T. *Environ* 2000, 78:167-174.
45. Medina, A., Probanza, A., Gutierrez Mañero, F.J. et al. *Appl Soil Ecol* 2003, 22:15-28.
46. Azcón-Aguilar, C., Palenzuela, J., Roldán, A. et al. *Appl Soil Ecol* 2003, 22:29-37.
47. Phipps, C.J. and Taylor, T.N. *Mycologia* 1996, 88:707-714.
48. Schüßler, A. *Mycorrhiza* 2000, 10(1):15-21.
49. Brundrett, M.C. *New Phytol* 2002, 154(2):275-304.
50. Taylor, T.N. and Taylor, E.L. Prentice Hall, New Jersey, 1993.
51. Simon, L.K., Bousquet, J., Levesque, R.C. et al. *Nature* 1993, 363(6424):67-69.
52. Khan, A.G. *In*: Proceedings of Croucher Foundation Study Institute: Wetland Ecosystems in Asia Function and Management. Wong, M.H. (ed.), Hong Kong, 2003, pp. 11-15.
53. Smith, S.S. and Read, D.J. *Mycorrhizal Symbiosis*, 2nd ed., Academic Press, London, 1997.
54. Chaudhry, M.S., Batool, Z. and Khan, A.G. *Mycorrhiza* 2005, 15(8):606-611.
55. Khan, A.G. *In*: The Restoration and Management of deserted Land: Modern Approaches. Wong, M.H. and Bradshaw, A.D. (eds), World Scientific Publishing, Singapore, 2002, pp. 80-92.
56. Boruvka, L. and Drabek, O. *Plant Soil Environ* 2004, 50:339-345.
57. Kapoor, A. and Viraraghavan, T. *Biores Technol* 1995, 53(3):195-206.
58. Morley, G.F. and Gadd, G.M. *Mycol Res* 1995, 99:1429-1438.
59. Gonzalez-Chavez, M.C., Carrillo-Gonzalez, R., Wright, S.F. et al. *Environ Pollution* 2004, 130(3):317-323.
60. Wright, S.F. and Upadhyaya, A. *Plant and Soil* 1998, 198(1):97-107.
61. Driver, J.D., Holben, W.E. and Rilling, M.C. *Soil Biol Biochem* 2005, 37(1):101-106.
62. Dodd, J.C., Boddington, C.L., Rodríguez, A. et al. *Plant and Soil* 2000, 226(2):131-151.
63. Kaldorf, M., Kuhn, A.J., Schroder, W.H. et al. *J Plant Physiol* 1999, 154:195-206.
64. Gonzalez-Chavez, C., Dhaen, J., Vangronsveld, J.J. et al. *Plant and Soil* 2002, 240(2):287-297.
65. Joner, E.J., Briones, R. and Leyval, C. *Plant and Soil* 2000, 226(2):227-234.
66. Rillig, M.C., Ramsey, P.W., Morris, S. et al. *Plant and Soil* 2003, 253(2):293-299.
67. Khan, A.G. *J Trace Elem Med Biol* 2005, 18(4):355-364.
68. Li, X.L. and Christie, P. *Chemosphere* 2000, 42:201-207.
69. Joner, E.J., Leyval, C. and Briones, R. *Biol Fertil Soils* 2000, 226:227-234.

70. Tullio, M., Pierandrei, F., Salerno, A. et al. *Biol Fertil Soils* 2003, 37:211-214.
71. Turnau, K., Kottke, I. and Oberwinkler, F. *New Phytol* 1993, 123:313-324.
72. Zhou, J.L. *Appl Microbiol Biotechnol* 1999, 51:686-693.
73. Joner, E.J., Ravnskov, S. and Jakobsen I. *Biotechnol Lett* 2000, 22:1705-1708.
74. Li, X.L., George, E. and Marschner, H. *New Phytol* 1991, 119:397-404.
75. Olsson, P.A., Francis, R., Read, D.J. and Söderström, B. *Plant Soil* 1998, 201:9-16.
76. Laheurte, F., Leyval, C. and Berthelin, J. *Symbiosis* 1990, 9:111-116.
77. Friedel, J.K., Langer, T., Siebe, C. et al. *Biol Fert Soils* 2000, 31:414-421.
78. Christensen, J.S. and Elton, J. Groundwater pollution primer CE 4594: Soil and groundwater pollution. Civil Engineering Department, Virginia Technology, 1996.
79. Iturbe, R., Flores, R.M. and Torres, L.G. *Environ Monit Assess* 2004, 91:237-255.
80. Han, M.J., Choi, H.T. and Song, H.G. *J Microbiol* 2004, 42(2):94-98.
81. Bianciotto, V., Bandi, C., Minerdi, D. et al. *Appl Environ Microbiol* 1996, 62:3005-3010.
82. Xiaolin, L., Chen, B. and Feng, G. et al. Acts of the 17th World Congress of Soil Science. Symposium No. 42, Paper No. 1649, 2002.
83. Joner, E.J. and Leyval, C. *Agronomie* 2003, 23:495-502.
84. Cabello, M.N. *FEMS Microbiol Ecol* 1997, 22:233-236.
85. Huang, H.L., Zhang, S.Z., Shan, X.Q. et al. *Environmental Pollution* 2007, 146:452-457.
86. Wu, N.Y., Zhang, S.Z., Huang, H.L. et al. *Science of the Total Environment* 2008, 394:230-236.
87. Wu, N.Y., Zhang, S.Z., Huang, H.L. et al. *Environmental Pollution* 2008, 151:569-575.
88. Purin, S. and Rillig, M.C. *FEMS Microbiology Letters* 2008, 279:8-14.
89. Joner, E.J., Johansen, A., dela Cruz, M.A.T. et al. *Environmental Science and Technology* 2001, 35:2773-2777.
91. Lopez-Gutierrez, J.C., Toro, M. and Lopez-Hernandez, D. *Ecosystems Environment* 2004,103:405-411.
92. Huang, H.L., Zhang, S.Z., Chen, B.D. et al. *J Agric Food Chem* 2006, 54:9377-9382.
93. Memon, A.R., Aktoprakligil, D., Ozdemir, A. et al. *Soil Sci Plant Nut* 2001, 25: 111-121.
94. Joner, E. and Levylal, C. *New Phytol* 1997, 135:353-360.
95. Weissenhorn, I., Leyval, C. and Berthelin, J. *Biol Fertil Soil* 1995, 19:22-28.
96. del Val, C., Barea, J.M. and Azcòn-Aguilar, C. *Appl Environ Microbiol* 1999, 99:718-723.
97. Leyval, C., Singh, B.R. and Joner, E.J. *Water Air Soil Pollut* 1995, 84:203-216.
98. Weissenhorn, I., Leyval, C. and Berthelin, J. *Plant Soil* 1993, 157:247-256.
99. Gildon, A. and Tinker, P.B. *New Phytol* 1981, 95:263-268.
100. Gildon, A. and Tinker P.B. *Trans Br Mycol Soc* 1983, 77:648-649.
101. Sambandan, K., Kannan, K. and Raman, N. *J Environ Biol* 1992, 13:159-167.
102. Turnau, K., Ryszka, P., Gianinazzi-Pearson, V. et al. *Mycorrhiza* 2001, 10:169-174.
103. Griffioen, W.A.J., Iestwaart, J.H. and Ernst, W.H.O. *Plant Soil* 1994, 158:83-89.

104. Weissenhorn, I., Glashoff, A., Leyval, C. and Berthelin, J. *Plant Soil* 1994, 167:189-196.
105. Tonin, C., Vandenkoornhuyse, P., Joner, E.J. et al. *Mycorrhiza* 2001, 10:161-168.
106. Hildebrandt, U., Kaldorf, M. and Bothe, H. *J Plant Physiol* 1999, 154:709-717.
107. Shetty, K.G., Hetrick, B.A. and Schwab, A.P. *Environ Pollut* 1995, 88:307-314.
108. Arriagada, C.A., Herrera, M.A. and Ocampo, J.A. *Water Air and Soil Pollution*, 2005, 166:31-47.
109. Rabie, G.H. *Mycobiol* 2005, 33:41-50.
110. Blaylock, M.K., Salt, D.E., Dushenkov, S. et al. *Environ Sci Technol* 1997, 31(3):860-865.
111. Khan, A.G., Kuck, C. and Chandhry, T.M. *Chemosphere* 2000, 41(1-2):197-207.
112. Harrison, M.J. *Plant Soil* 1998, 184:195-205.
113. Awotoye, O.O. *Environtropica* 2006, 3(1&2):72-82.
114. Diaz, G., Azcón-Aguilar, C. and Honrubia, M. *Plant Soil* 1996, 180:241-249.
115. Turnau, K. and Mesjasz-Przybylowicz, J. *Mycorrhiza* 2003, 13:185-190.
116. Tullio, M., Pierandrei, F., Salerno, A. et al. *Biol Fertil Soils* 2003, 37:211-214.
117. Vivas, A., Marulanda, A., Gómez, M. et al. *Soil Biol Biochem* 2003, 35:987-996.
118. Kaldorf, M., Kuhn, A.J., Schroder, W.H. et al. *J Plant Physiol* 1999, 154:718-728.
119. Tonin, C., Vandenkoornhuyse, P., Joner, E.J. et al. *Mycorrhiza* 2001, 10:161-168.

8

Rhizosphere Management Practices in Sustaining Mycorrhizae

C. Harisudan, M. Velmurugan[1] and P. Hemalatha

Directorate of Extension Education, Tamil Nadu Agricultural
University, Coimbatore 641003, Tamil Nadu, India
[1]Horticultural College and Research Institute, Tamil Nadu Agricultural
University, Coimbatore 641003, Tamil Nadu, India

Introduction

Soil is a life supporting system, a natural and a vital resource for growing food, fibre and firewood to meet the human needs. The importance of soil biota was realized by the agricultural scientist and farmers recently. Since, Hiltner defined rhizosphere in 1904 some remarkable advances have been made in recognition of the important role of microorganisms (and to some extent micro fauna) in the rhizosphere soil. According to Arshad and Frankenberger [1] rhizosphere is the portion of the soil under the direct influence of the roots of higher plants. It is considered the most intense ecological habitat in soil in which microorganisms are in direct contact with plant and it is associated with a distinct, diverse community of metabolically active soil microbiota that carry out biochemical transformations.

Rhizosphere is defined as the soil adjacent to roots with a different physical, chemical and biological environment from the bulk soil [2]. The rhizoplane is defined as the actual root surface soil interface. The soil, roots and the microorganism are the three major components of rhizosphere which decides the success of crop production. It is probable that the best management techniques will involve manipulation of all the three components. The increasing interest in using microorganisms (bioaugmentation) to achieve the dream of low input sustainable agriculture and to circumvent expensive and possibly environmentally deleterious agricultural chemicals focuses new attention on more effective management of the rhizosphere. Soil microorganisms play a pivotal role both in the evolution of agriculturally useful soil conditions and in stimulating plant growth.

Biological activities essentially control soil productivity. Soil invertebrates, fungi and bacteria recycle all the carbon, nitrogen and other mineral nutrients in plant and animal residues into forms that can be used by plants. Soil microbes (e.g. bacteria and fungi) regulate the destruction of toxic environment pollutants like nitrous oxide and methane. Increasing the diversity of the soil will increase the ability of the soil to adapt and tolerate a variety of environmental conditions. The speed at which residues decay and nutrients are released from organic matter will largely depend on how we manage the soil. The interface between soil and plant roots, the rhizosphere is dynamic habitat. In the surrounding bulk soil, the growth and proliferation of microbes is normally limited, the plant rhizosphere facilitates the activity and multiplication of large variety of microorganisms. Different plant species release different organic compounds, and thus, they can encourage a differentiated rate of proliferation of the microbiota [3]. In this era sustainable technologies need to be adopted in farming to increase the concern for environmental quality. Adopting proper rhizosphere management practices in sustaining mycorrhizae is one of the important approaches. The role of mycorhizae in plant is obvious.

Mycorrhizae and Rhizosphere Environment

Mycorrizhae is an association between plants and fungi that colonize the cortical tissues of roots during periods of active plant growth [4]. Several research findings have reported about the positive growth responses of plants to mycorrhizal infection, particularly on phosphorus deficient soils [5]. This has been attributed to increased P uptake via the fungal symbionts which in turn derive carbon compounds from the autotrophic hosts. The presence of mycorrhizae in rhizosphere environment improves the soil physical property through the mycelia strands produced by them. The best aggregating effect on the soil is also obtained by the fungal hyphae. The aggregating power of micro-organism is dependent on the organic matter content of the soil. The soil, plant roots and the microorganism are the three major components of rhizosphere which play a vital role in success of crop production. It is probable that the best management techniques will involve manipulation of all the three components.

Rhizosphere Management Techniques

The management options embrace all the three components. Even if one component is affected due to improper/faulty management practices, the other two will also be affected. Hence all the three components should be managed properly.

Soil Manipulation

Soil disturbance or tillage operations play a vital role in the manipulation of rhizosphere components viz., soil, root and microorganisms. Tillage improves physical conditions of soil that results in better water retention, aeration, temperature, and reduces erosion. Manipulating the soil physical properties provides favourable environment for the root respiration and root growth penetration.

Tillage Impact on Rhizosphere Microorganism and Mycorrhizae

Soil disturbance influences rhizosphere microorganisms; the negative effect on AM fungi thereby reduces the benefits to crops and soil productivity. Tillage operations provide both negative and positive effect. It cuts the disease cycle particularly the root disease; at the same time the beneficial microbes are affected by the tillage operations. Conservation tillage practices create a more undisturbed rhizosphere environment. Colonization by AM fungi is reduced under conventional tillage compared with no tillage. Soil manipulation also reduces the inoculation potential of the soil and the efficacy of mycorrhizas by disrupting the extraradical hyphal network [6, 7]. The hyphal network is rendered non-infective by tillage [6]. The disruption of the hyphal network decreases the absorptive abilities of the mycorrhizae because the surface area spanned by the hyphae is greatly reduced, which result in lower phosphorus input to the plants which are connected to the hyphal network [6]. Heavy phosphorus fertilizer input may not be required in reduced tillage system as compared to heavy tillage systems, which might be due to the increase in mycorrhizal network which allows mycorrhizae to provide the plant with sufficient phosphorus.

Many research findings have reported about the negative impact of tillage. Mulligan et al. [8] proposed that the negative impact of tillage on root colonization was due to lower root growth of dry bean (*Phaseolus vulgaris* L.) caused by increased soil bulk densities in tilled soils. Schenk et al. [9] observed an increase in mycorrhizal spores and root colonization with minimum tillage compared with conventional tillage. But, O'Halloran et al. [10] observed greater root growth but lower P uptake of corn, suggesting that the negative effects of tillage were not the result of reduced root growth. O'Halloran et al. [10] reported that soil disturbance decreases early plant growth, P uptake and AM colonization in corn. Furthermore, when a non-mycorrhizal canola plant [10] or spinach [11] was grown, soil disturbance did not have any effect on P absorption of those plants. These results indicate that the negative effect of disturbance on P uptake is likely due to impaired AM association. Sylvia [12] reported that extraradical hyphae might be the main source of inoculum in soil especially when host plants are present and soil is not tilled for crop production. Evans and Miller [13] observed that disturbance of root-free soil containing only AM hyphae detached from host plants reduced the AM colonization of corn roots planted later in this soil, and decreased plant growth and nutrient uptake. This suggested that if the AM hyphal network is not disrupted, the next crop will be more rapidly connected to the network and nutrient absorption capacity would be enhanced. Jasper et al. [14] in Australia found a reduction of AM colonization of clover after soil disturbance and suggested that most of this reduction was due to decreased hyphal viability. On the contrary [15] a study in eastern Canada on the effect of soil disturbance on AM colonization and corn growth, report that decrease in P uptake and plant growth were not accompanied by a decrease in AM colonization. McGonigle et al. [15] proposed that if AM fungi were an important component of the disturbance effect, it would have to be through the dismantling of a potential though dynamic hyphal network rather than through reduction in the mycorrhizal colonization potential of the soil. The role of extraradical hyphae as principal propagules for AM colonization might be of considerable importance, particularly in cool climates where population of the viable spores in agricultural soils may be extremely low following winter [16]. It is evidenced from the recent research studies that AM were involved in P uptake and were negatively affected by soil disturbance.

Soil Moisture Impact on Rhizosphere Component

Soil moisture stress will affect survival of different types of propagules differently. Generally, thin-walled hyphae will be most susceptible to drying while thick-walled e.g., *R. solani* and thick-walled spores/cysts and specialized survival structures such as sclerotia are the least susceptible. However, survival/ moisture interactions are also complicated by the effects of moisture on the activities of microbial antagonists and the effect of wetting and drying on resistance to antagonist. Allen [17] found that increased moisture tended to increase the VAM length. It is suggested that timely and optimum irrigation is essential. Research findings have revealed greater survival of mycorrhizal seedling than non-mycorrhizal seedling under drought condition [18]. The consumptive use of water and water use efficiency of greengram is highest under VAM treated plot [19].

Effect of Crop Rotation on the Rhizosphere Environment

The repetitive cultivation of an ordered succession of crops or crops and fallow on the same land is called crop rotation. The prime energy source for microbial growth in the soil is particulate plant organic matter, much of which is of root origin. Organisms colonizing the rhizosphere of one crop and surviving in these root residues will provide a significant potential source of inoculum for successive crops. Continuous monoculture cropping can act as an enrichment medium for rhizosphere microorganisms that proliferate on the roots of the particular monoculture crop. The best evidence for this is the build-up of soil-borne root diseases when susceptible crops are grown and in the creation of disease-suppressive soils with continuous monoculture in the presence of root disease.

It is postulated that the rate of change of the composition of rhizosphere microflora on changing the plant species will be related to root biomass; this may be one reason why the build-up of specific suppression to disease in soil is faster under the intensive production systems. The management of alternate crops is also an important strategy in the management of soil populations of target organisms. Microbial biomass C was highest in pearl millet-wheat-green manure rotation and was lowest in pearl millet-wheat rotation [20]. With organisms having a positive effect on plant growth, the choice of rotations to enhance the population is important and it should be just feasible for agronomist to devise rotations to enhance the population of AM mycorrhizal fungi.

Bioaugmentation

It is increasing the activity of beneficial microorganisms in rhizosphere by adding microorganism to the environment, a technique used for enhancing crop productivity. Bioaugmentation is defined as the addition of pre-grown microbial cultures to perform a remediation task in a given environment.

Under circumstances where native mycorrhizal inoculum potential is low or ineffective, addition of pre-grown appropriate fungi for the plant production system is effective. Crop management strategies alone cannot always sustain efficient AM fungal communities, it would be useful for agriculturists to have access to efficient, economical and easy-to-use AM inoculants [21]. The first step in any inoculation programme is to obtain an isolate that is both infective, and able to penetrate and spread in the root, and effective or able to enhance growth or stress tolerance of the host. Isolation and inoculum production of EM and AM fungi present very different problems. Many EM fungi can be cultured on artificial media. Therefore, isolates of EM fungi can be obtained by placing surface disinfected portions of sporocarps or mycorrhizal short roots on an agar growth medium. The resulting fungal biomass can be used directly as inoculum but, for ease of use, inoculum often consists of the fungal material mixed with a carrier or bulking material such as peat. The biotrophic (obligate) nature of AM fungi has traditionally hampered inoculum production and, therefore, obtaining isolates of AM fungi is more difficult because they will not grow apart from their host [22]. Culture of AM fungi on plants growing in disinfected soil has been the most frequently used technique for increasing propagule numbers [23]. But, new techniques aimed at the production of AM fungi [24] and formulation of inoculants [25] could foster the widespread practice of crop inoculation in the near future. Spores can be sieved from soil, surface disinfected and, used to initiate 'pot culture' on a susceptible host plant in sterile soil or an artificial plant growth medium [22]. When the required species multiply on the roots, the authenticity of the species must be checked microscopically. Once this is assured, the entire root system is finely chopped with adhering soil and used as the inoculum for scaling up inoculum production on living roots of test host grown in pots containing sterilized soil. Inoculum is typically produced in scaled up pot cultures. Onions, sorghum and other grasses are suitable hosts to multiply AM fungal inocula and the plants are grown in 1:1 sand soil mixture to obtain good root growth.

Hydrophonic and aeroponic culture systems are also possible. A benefit of these systems is that plants can be grown without a supporting substratum, allowing colonized roots to be sheared into an inoculum of high propagule number. The following points should be considered for inocula production. A highly susceptible host plant should be used and it should produce root mass quickly and tolerate high light conditions required for the fungus to reproduce rapidly [26]. Host plants should be screened to ensure that maximum inoculum levels are achieved [27]. Seed propagated hosts are preferable to cuttings since seeds can be more easily disinfected than cuttings [26]. All components of the culture system should be disinfected prior to initiation of a culture of AM fungi. The objective of soil disinfection is to kill existing AM fungi, pathogenic organisms and weed seeds while preserving a portion of non-pathogenic microbial community [26]. Containers used for the pot cultures should be shielded from contaminated soil, splashing water and crawling insects. In addition, specific isolates of AM fungi should be kept well separated to reduce cross contamination. Container size should match the potential volumes of the root system within practical space constraints. Larger containers may result in higher spore concentration [28]. Light quality and intensity and soil moisture and temperature markedly influence root colonization and spore production [23]. A moderate irrigation regime [29] and a warm growth environment [30] support optimal colonization and spore production. Fungal responses to P and N fertilization are strain dependent. A nutrient regime low in P but high in N increases colonization of AM fungi [31]. Some soil-applied biocides greatly reduce or eliminate AM fungi [32], while others increase colonization and sporulation [33]. Selected use of pesticides may be useful for better inoculum production, but they are rarely used to protect against or reduce organisms contaminating pot cultures [34].

Field experiments in turmeric (*Curcuma longa*) reveal that the combined application of FYM and bioaugmentation of azospirillum, phosphobacteria and VAM exhibited higher rhizome yield (33313.00 kg ha^{-1}) as compared to the application of 50 per cent recommended dose of fertilizer + phosphobacteria (18636.00 kg ha^{-1}). Moreover the nutrient content and nutrient uptake of turmeric was also enhanced by the bioaugmentation and FYM application [35-39].

Beneficial Ectomycorrhizae Species

Ectomycorrhiza is an association of the fungus and the feeder roots (root hairs) in which the fungus grows predominantly intercellularly in the cortical region penetrating the epidermis by secreting proteolytic enzymes and develops extensively outside the root forming a network of hyphae which is known as 'Hartig net'. Ectomycorrhiza involve an estimated 5000 species of fungi belonging to the subphyla Basidiomycetes and Ascomycotina and some 2000 species of higher plants, gymnosperms and angiosperms, both monocots and dicots of different families such as Pinaceae, Fagaecae and Myrtaceae etc. The mycorrhizal fungus basically serves as an extension of the plants root systems, exploring soil far beyond the reach of the plants root systems, and transporting water and nutrients to the roots. The uptake of phosphorus and nitrogen are especially critical functions of mycorrhizal fungi, it can release bound form of these nutrients otherwise unavailable to the roots. Some of the beneficial ectomycorrhizae are *Pisolithus tinctorius, Rhizopogon fulvigleba, Rhizopogon rubescens, Rhizopogon villosulus* and *Rhizopogon vulgaris*.

Beneficial Arbuscular (Endomycorrhizae) Species

Endomycorrhizae do not form a mantle over the root but the fungus actually enters the cortex cells. It grows predominantly within the root cells. The hyphae form coils and swellings that eventually disappear as a result of digestion of the invaded cells and the hypha follows the advancing meristematic root tips while its mycelium extends far away in soil and thus increases the surface area for absorption of nutrients. Most of the endomycorrhiza commonly occur in various economically important plants and are characterized as Vesicular-Arbuscular Mycorrhiza (VAM) recognized by the presence of 'vesicles' (terminal spherical structures containing oil droplets) and 'arbuscles' (complex structures formed by repeated dichotomous branching of hyphae in cortical cells of the feeder roots). They function as reserve organs and also for fungal multiplication. Arbuscles absorb nutrients from the plant cells but also release mineral elements to the plant. Some of the important genera of endomycorrhizae are *Glomus aggregatum, G. mosseae, G. intraradices* etc.

Benefits of Bioaugmentaion in the Rhizosphere

Microorganisms contribute a wide range of essential services to the sustainable functioning of all ecosystems. Various soil microorganisms that are capable of ever thing beneficial effects on plant pests and diseases either in culture or protected environment have potential for in agriculture and lead to increased yields of a wide variety of crops. The bioaugmentation plays a multibeneficiary role viz., nutrient cycling, regulating the dynamic of soil organic matter, soil carbon sequestration, modifying soil physical structure and water regimes, enhancing the amount and efficiency of nutrient acquisition by the vegetation, controlling diseases and insect pests, improving the drought tolerance of plants, phosphate solubilizers and mobilizers, production of plant growth regulators, siderophores, antibiotics and degrading synthetic soil contaminants and other potential pollutants.

Nutrient Dynamics

The benefit of mycorrhizae to plants is mainly attributed to increased uptake of nutrients, especially phosphorus. This increase in uptake may be due to increased surface area of soil contact, increased movement of nutrients into mycorrhizae, a modification of the root environment and increased storage [40]. Mycorrhizae can be much more efficient than plant roots at taking up phosphorus. Phosphorus travels to the root or via diffusion and hyphae reduce the distance required for diffusion thus increasing uptake. The rate of inflow of phosphorus into mycorrhizae can be up to six times that of the root hairs [40]. In some cases the role of phosphorus uptake can be completely taken over by the mycorrhizal network and all of the plant's phosphorus may be of hyphal origin [41].

The available phosphorus concentration in the root zone can be increased by mycorrhizal activity. Mycorrhizae lower the rhizosphere pH due to selective uptake of NH_4^+ (ammonium ions) and release of H^+ ions. Decreased soil pH increases the solubility of phosphorus precipitates. The hyphal uptake of NH_4^+ also increases the flow of nitrogen to the plant as NH_4^+ is adsorbed to the soil's inner surfaces and must be taken up by diffusion.

Soil Quality

Sustaining native AM fungi increase the success of ecological restoration project and the rapidity of soil recovery [42]. The production of a soil protein known as glomalin enhances soil aggregate stability. Glomalin improves soil aggregate water stability and decrease soil erosion. A strong correlation has been found between GRSP and soil aggregate water stability in a wide variety of soils where organic material is the main binding agent, although the mechanism is not known. The protein glomalin has not yet been isolated and described, and the links between glomalin, GRSP and arbuscular mycorrhizal fungi is not yet clear [43].

Effect of VAM Fungi on Nodule Number, Nitrogenase Activity and Rhizosphere Microflora of Chickpea

Rakesh Kumar et al. [44] observed significant increase in nodule number and nitrogenase activity in chickpea. Rhizosphere fungal population significantly reduced by inoculation with different VAM fungi, while bacterial and Azotobacter population significantly increased with the mycorrhizal inoculation (Table 1). Many researches have proven that AM fungi release an unidentified diffusional factor, known as the myc factor, which activates the nodulation factor's inducible gene mtENOD11. This is the same gene involved in establishing symbiosis with the nitrogen fixing bacteria, rhizobium [45]. When rhizobium bacteria are present in the soil, mycorrhizal colonization is increased due to an increase in the concentration of chemical signals involved in the establishment of symbiosis [46].

Mycorrhizal plants produced higher fruit yields (Table 2) under severe, moderate, mild drought stressed and well watered conditions than non-mycorrhizal plants. The results suggest that mycorrhizal colonization affects host plant nutritional status, water status and growth and thereby alters reproductive behaviour, fruit production and quality of fruits regardless of well-watered or drought-stressed conditions. Further, there is a possibility of saving of fertilizer N or P to the tune of 30-40% in nutrient deficient soils [47].

Table 1. Effect of VAM fungi on nodule number, nitrogenase activity and rhizosphere microflora of chickpea [44]

Parameters	Glomus mosseae	Control
Nodule number/plant	17.3	15.2
Nitrogenase activity (μ mole/h/g nodule weight)	4.16	3.27
Fungal CFU/g soil $\times 10^4$	9.67	15.66
Actinomycetes CFU/g soil $\times 10^5$	13.83	11.67
Bacteria CFU/g soil $\times 10^6$	12.00	4.16
Azotobacter CFU/g soil $\times 10^4$	9.70	4.32

Table 2. Performance of mycorrhizal and non-mycorrhizal tomato plants [47]

Drought stress	Yield (t/ha)	
	Mycorrhizal	Non-mycorrhizal
Severe	15.8	11.8
Moderate	17.2	13.3
Mild	19.2	15.9
No stress	20.3	17.5

Conclusion

Soil is a pot full of weathered geological ingredients that have been stewed together through time with billions of tiny living creatures, viz. plant roots, viruses, bacteria, fungi, algae, protozoa, mites, nematodes, worms and other fauna. Managing or manipulating the rhizosphere component is essential to enhance the yield of agricultural crops. The soil applied cultures maintain an optimum population, to suppress the other pathogenic microbes; the population of applied beneficial microorganism should be increased. Hence through agronomic practices we are able to improve the population of rhizosphere microorganism.

Mycorrhizal systems must be considered as an essential factor for soil fertility and for promoting plant health and productivity. It is well-documented that mycorrhizal fungi can transfer phosphorous to the host and that results in an increase of P uptake by plants colonized by these fungi. The benefits of high fertilization rates on yield are well established and very few producers are ready to reduce their P-fertilization regime to levels required to allow adequate development of the symbiosis. It must also be recognized that recent research on the possible application of the mycorrhizal symbiosis in agriculture has revealed many gaps in our knowledge of fungal biology and ecology. Therefore, before the best use of these symbioses can be made in crop production, the effect of agronomic practices on indigenous communities of mycorrhizal fungi must be well elucidated. Efficient fungal isolates could be selected for various desirable traits, multiplied and inoculated onto crops, but the conditions which would allow these organisms to express their full potential will have to be determined. A better understanding of the conditions required for utilization of AM inocula will improve inoculation of field crops and lead to increased numbers and quality of fungi in agricultural soils. Choice of suitable crop in crop rotations, suitable crop in intercropping, practicing organic farming, reducing the usage of agrochemicals, maintaining optimum soil moisture and minimum tillage maintains optimum population of mycorrhizae. It could be concluded that managing the rhizosphere ecosystem, controlling rhizosphere processes and sustaining mycorrhizae towards sustainable development can be one of the most important ways to enhance nutrient-resource utilization efficiency and enhance crop productivity.

References

1. Arshad, M. and Frankenberger, W.T. *Adv. Agron* 1998, 62:45-151.
2. Wallace, J. Canadian Organic Growers Inc., Mothersill Printing, Canada, 1988.
3. Olsson, S. and Alstrom, S. *Soil Biol. Biochem* 2000, 32:1443-1451.
4. Miller, R.M. and Jastrow, J.D. *In:* Mycorrhizae and Plant Health. Pfleger, F.L. and Linderman, R.G. (eds), APS Press, St. Paul, Minnesota, pp. 189-212.
5. Son, C.L. and Smith, S.E. *New Phytologist* 1988, 108:305-314.
6. Mc Gonigle, T.P. and Miller, M.H. *Appl. Soil Ecol* 1999, 12:41-50.
7. Mozafar, A., Anken, T., Ruh, R. et al. *Agron. J* 2000, 92:1117-1124.
8. Mulligan, M.F., Smucker, A.J.M. and Safir, G.F. *Agron. J* 1985, 77:140-144.
9. Schenk, N.C., Smith, G.S., Mitchell, D.J. et al. *Florida Sc.* 1982, 45(S):8.

10. O'Halloran, I.P., Miller, M.H. and Arnold, G. *Can. J. Soil Sc* 1986, 66:287-302.

11. Evans, D.G. and Miller, M.H. *New Phytol* 1988, 110:67-74.

12. Sylvia, D.M. *In:* Methods in microbiology, Vol. 24. Norris, J.R., Read, D.J. and Varma, A.K. (eds), Academic Press, London, 1992, pp. 53-65.

13. Evans, D.G. and Miller, M.H. *New Phytol* 1990, 114:65-71.

14. Jasper, D.A., Abbott, L.K. and Robson, A.D. *New Phytol* 1989, 112:93-99.

15. Mc Gonigle, T.P., Evans, D.G. and Miller, M.H. *New Phytol* 1990, 116:629-636.

16. Addy, H.D., Miller, M.H. and Peterson, R.L. *New Phytol* 1997, 135:745-753.

17. Allen, M.F., Richards, J.H. and Busso, C.A. *Biol. Fertil. Soils* 1989, 8(4):285-289.

18. Peter, D.S., Sandra, M.F. and Stephen, E.W. *Soil Sci Soc Am J* 1988, 62:1309-1313.

19. Idnani, I.K. and Singh, R.J. *Indian J. Agrl. Sci.* 2008, 78(1):53-57.

20. Dhull, S.K., Goyal, S., Kapoor, K.K. et al. *Indian J. Agric. Res.* 2005, 39(2):128-132.

21. Hamel, C. *Agriculture, Ecosystem and Environment* 1996, 60:197-210.

22. Sylvia, D.M. *In:* Methods of Soil Analysis, Part 2, Microbiological and Biochemical Properties. Weaver, R.W. (ed.), Soil Science Society of America, Madison, WI, 1994, pp. 351-378.

23. Menge, J.A. VA Mycorrhiza. CRC Press, Inc., Boca Raton, Florida, 1984, pp. 187-203.

24. Becard, G. and Fortin, J.A. *New Phytologist* 1988, 108:211-218.

25. Strullu, D.G. and Plenchette, C. *Mycology Research* 1991, 95:1194-1196.

26. Sylvia, D.M. and Jarstfer, A.G. *In:* Management of Mycorrhizas in Agriculture, Horticulture and Forestry. Robson, A.D., Abbott, L.K. and Malajczuk, N. (eds), Kluwer Academic Publisher, Netherlands, 1994, pp. 231-238.

27. Bagyaraj, D.J. and Manjunath, A. *Plant Soil* 1980, 55:495-498.

28. Ferguson, J.J. and Menge, J.A. *Proceeding of Florida State Horticultural Society* 1982, 95:37-39.

29. Nelson, C.E. and Safir, G.R. *Planta* 1982, 154:407-413.

30. Schenck, N.C. and Smith, G.S. *New Phytologist* 1982, 92:193-201.

31. Sylvia, D.M. and Neal, L.H. *New Phytologist* 1990, 115:303-310.

32. Trappe, J.M. *In:* Ecophysiology of VA Mycorrhizal Plants. Safir, G. (ed.), CRC Press, Boca Raton, USA, 1987, pp. 5-25.

33. Sreenivasa, M.N. and Bagyaraj, D.J. *Plant and Soil* 1989, 119:127-132.

34. Menge, J.A. *Canadian Journal of Botany* 1983, 61:1015-1024.

35. Velmurugan, M., Chezhiyan, N. and Jawaharlal, M. *Asian Journal of Soil Science* 2007, 2(2):113-117.

36. Velmurugan, M., Chezhiyan, N. and Jawaharlal, M. *Asian Journal of Horticulture* 2007, 2(2):23-29.

37. Velmurugan, M., Chezhiyan, N., Jawaharlal, M. et al. *Asian Journal of Horticulture* 2007, 2(2):291-293.

38. Velmurugan, M. and Chezhiyan, N. *South Indian Horticulture* 2007, 53(1-6):392-396.

39. Velmurugan, M., Davamani, V., Anand, M. et al. *Plant Archives* 2008, 8(1):9-32.

40. Bolan, N.S. *Plant and Soil* 1991, 134:189-207.

41. Smith, S., Smith, A. and Jakobsen, I. *Plant Physiology* 2003, 133:16-20.
42. Rillig, M., Ramsey, P., Morris, S. et al. *Plant and Soil* 2003, 253:293-299.
43. Rillig, M. *Canadian Journal of Soil Science* 2004, 84:355-363.
44. Kumar, B.L.R. and Jalali, C.H. *Legume Res.* 2004, 27(1):50-53.
45. Kosuta, S., Chabaud, M., Lougnon, G. et al. *Plant Physiology* 2003, 131:952-962.
46. Xie, Z., Staehelin, C., Vierheilig, H. et al. *Plant Physiology* 1995, 108:1519-1525.
47. Subramanian, K.S. and Santhanakrishnan, P. *J. Agril., Res. Mgt.* 2005, 4(S):287-288.

9

Importance of Mycorrhizae for Horticultural Crops

**P. Hemalatha, M. Velmurugan[1], C. Harisudan
and V. Davamani[2]**

Directorate of Extension Education, Tamil Nadu Agricultural
University, Coimbatore 641003, Tamil Nadu, India

[1]Horticultural College and Research Institute, Tamil Nadu Agricultural
University, Coimbatore 641003, Tamil Nadu, India

[2]Agro Climate Research Centre, Tamil Nadu Agricultural
University, Coimbatore 641003, Tamil Nadu, India

Introduction

During the past several decades, production of horticultural crops has undergone a tremendous change in both developed and developing countries due to the development and use of fertilizers, pesticides, mechanization, soil fumigation, container production in soil-less media, breeding, micropropagation, pathogen elimination and a number of other innovations for attaining its future stability and sustainability. Recently, in modern agriculture, enhanced productivity depends heavily on inputs of fertilizers and pesticides to maintain an economic level of productivity. On the other hand, social and economic pressures are demanding that agricultural practices, that have a damaging impact on the environment and pose risk to the health of mankind, be curtailed. In the near future, we would face severe problem in fertilizer production which is an important factor for attaining technological revolution. The production of nitrogen fertilizers seems to be energy intensive and reserves of some fertilizer components especially phosphate are becoming limiting factors. Thus throughout the world, agriculture including horticulture is at crossroads in terms of its future stability and sustainability. The solution for the above lies in developing new technology with reduced use of fertilizer and pesticide besides maintaining a sustainable level of productivity.

The increased concern about the sustainability and undesirable side-effects of chemistry-based agriculture is prompting many researchers to more seriously evaluate alternative approaches. One such effort was put forth in the management of microbes that live in the rhizosphere soil around the roots of nearly all terrestrial plants. In recent years, a wide demonstration has been given about the beneficial or essential role of mycorrhizal fungi in the lives of many plants of horticultural interest, especially when growing in natural, unmanaged conditions. The organic cultivation techniques that include the use of manures, crop rotation, green manure and green leaf manure crops and bio-dyanamic farm approaches from decades ago were found to be effective mainly because they had influenced mycorrhizal fungi.

It has been believed that the main orchestrators of microbial activity in the rhizosphere are mycorrhizal fungi. Therefore, the successful management of the rhizosphere microflora depends on management of mycorrhizae, the symbiotic relationship that forms with special soil fungi and most plant roots. It is not proposed to discard the technical advances that led to our present state of horticultural crop production. However, in order to obtain the sustainable and environmentally compatible production at present, the role of mycorrhizal fungi is found to be very much significant.

Mycorrhizae

The word *Mycorrhizae* was first coined by German researcher A.B. Frank in 1885, and originates from the Greek *mycos*, meaning 'fungus' and *rhiza*, meaning 'root'. Mycorrhiza is a symbiotic mutalistic relationship between special soil fungi and fine plant roots. It is neither the fungus nor the root, but rather the structure formed from these two partners. This implies a true symbiotic relationship between fungi and plant roots, which is similar to that of nodule bacteria in legumes. The mutualistic association provides the fungus with relatively constant and direct access to mono- or dimeric carbohydrates, such as glucose and sucrose produced by the plant in photosynthesis [1]. The carbohydrates are translocated from their source location i.e., leaves to the root tissues and then to the fungal partners. In return, the plant gains the use of the mycelium's very large surface area to absorb water and mineral nutrients from the soil, thus improving the mineral absorption capabilities of the plant roots [2]. Plant roots alone may be incapable of taking up phosphate ions that are immobilized, for example, in soils with a basic pH. The mycelium of the mycorrhizal fungus can, however, access these phosphorus sources and make them available to the plants they colonize [3]. The mechanisms of increased absorption are both physical and chemical. Mycorrhizal mycelia are much smaller in diameter than the smallest root, and can explore a greater volume of soil, providing a larger surface area for absorption. Thus the fungus takes over the role of the plant's root hairs and acts as an extension of the root system. Mycorrhizae are present in 92% of plant families (80% of species) with vesicular arbuscular mycorrhizae being the ancestral and predominant form [4] and indeed the most prevalent symbiotic association found in all the plant kingdom [1].

Types of Mycorrhiza

The mycorrhiza is of three types i.e. (i) ectomycorrhiza, (ii) endomycorrhiza and (iii) ecto-endomycorrhiza. These three groups are differentiated by the fact that hyphae of ectomycorrhizal fungi do not penetrate individual cells within the root, while the hyphae of endomycorrhizal fungi penetrate the cell wall and invaginate the cell membrane and exhibiting both ecto and endomycorrhizal characteristics by ecto-endomycorrhizal fungi.

Ectomycorrhiza

Ectotrophic types, or ectomycorrhizae, are found in trees in the Pinaceae, Salicaceae, Tiliaceae families and in roots of some members of Rosaceae, Leguminosae, Ericaceae and Juglandaceae [5, 6]. This is an association of the fungus and the feeder roots (root hairs) in which the fungus grows predominantly intercellular in the cortical region penetrating the epidermis by secreting proteolytic enzymes and develops extensively outside the root forming a network of hyphae which is known as 'Hartig net'. Structurally, ectomycorrhizae can be distinguished by the presence of hyphal strands coalescing to form a thick sheath around the feeder roots known as mantle. The mantle is of variable thickness, colour and texture surrounding the rootlets. These hyphal strands are capable of permeating outward from the root surface several metres or more and exploring regions not accessible by root hairs. Ectomycorrhizal roots are generally recognizable by their short, swollen appearance and distinctive colours of white, black, orange, yellow or olive green depending on the colour of fungal symbionts. Ectomycorrhiza are also characterized by specific branching patterns ranging from monopodial to multi-forked or coralloid.

Ectomycorrhizal fungi can exist in the soil as spores, sclerotia and rhizomorphs [7]. Ectomycorrhiza absorbs and stores plant nutrients like nitrogen, phosphorus, potassium, calcium etc. in their mantle. Besides it also converts some organic molecules into simple, easily available forms. Most of the ectomycorrhizal fungi come in the class basidiomycetes and the common genera are *Amanita*, *Fuscoboletinus*, *Lecimum*, *Boletus*, *Cortinarious*, *Suillus*, *Pisolithus* and *Rhizopogan* etc. Besides the tuffle like fungi of the class Ascomycetes is also known to form ectomycorrhiza [8].

Endomycorrhiza

Endomycorrhizae are the most widely distributed of any of the mycorrhizae, being found in many herbaceous, shrub and tree species, including most annual horticultural crops [9]. These forms are known as Vesicular-Arbuscular Mycorrhizae. *Endotrophic* types, Endomycorrhizae or VAM do not form a mantle over the root and the fungus actually enters the cortex cells. It grows predominantly intracellular i.e. within the root cells. The hyphae form coils and swellings that eventually disappear as a result of digestion of the invaded cells and the hypha follows the advancing meristematic root tips while its mycelium extends far away in soil and thus increases the surface area for absorption of nutrients. Most of the endomycorrhiza commonly occurring in various economically important plants are characterized as VAM recognized by the presence of 'vesicles' (terminal spherical structures containing oil droplets) and 'arbuscles' (complex structures formed by repeated dichotomous branching of hyphae in cortical cells of the feeder roots) [8]. They may function as reserve organs [10] and also for fungal multiplication. Arbuscles absorb nutrients from the plant cells but also release mineral elements to the plant. VAM fungi also produce spores outside the root. Endomycorrhizal fungi have to be multiplied through a host plant. For this reason, widespread inoculations are difficult and better field success is obtained through management of the symbiosis (i.e., nutrient conditions of the soil/substrate of plant growth) [11]. The fungi forming endomycorrhiza mostly belong to the class zygomycetes and family endogonaceae. The important genera of such fungi include *Endogone*, *Gigaspora*, *Acculospora*, *Glomus*, *Solerocystis* etc.

Ecto-endomycorrhiza

As the name implies, this is the third type of mycorrhizae exhibiting both the ecto- and endomycorrhizal characteristics. This is an association of the fungus and the roots of plant representing a condition where typical ectotrophic intercellular infection is accompanied with intercellular penetration of hyphae. However, very little is known about the species of fungi involved. They appear to have a limited distribution in forest soils and have been associated with species in nursery seedbeds that are normally ectomycorrhizal [12].

VAM/AM Symbiosis

The diagnostic feature of Arbuscular Mycorrhizae (AM) is the development of a highly branched arbuscule within root cortical cells. The fungus initially grows between cortical cells, but soon penetrates the host cell wall and grows within the cell. AM is an ancient symbiosis that originated atleast 460 million years ago. Arbusscular Mycorrhizae is ubiquitous among land plants, which suggests that mycorrhizae were present in the early ancestors of existing land plants. This positive association with plants may have facilitated the development of land plants. The nature of the relationship between plants and the ancestors of Arbuscular Mycorrhizal fungi is contentious. Two hypotheses are that (1) mycorrhizal symbiosis evolved from a parasitic interaction which developed into a mutually beneficial relationship, and (2) mycorrhizal fungi developed from saprobic fungi that became endosymbiotic.

In both cases the symbiotic plant-fungi interaction is thought to have evolved from a relationship in which the fungi was taking nutrients from the plant into a symbiotic relationship where the plant and fungi exchange nutrients. The development of AM fungi prior to root colonization, known as presymbiosis, consists of following stages: spore germination, hyphal growth, host recognition and appressorium formation.

Once inside the parenchyma the fungi forms highly branched structures for nutrient exchange with the plant called 'arbuscules' [13]. These are the distinguishing structures of Arbuscular Mycorrhizal Fungus. Arbuscules are the sites of exchange for phosphorus, carbon, water and other nutrients. The host plant exerts a control over the intercellular hyphal proliferation and arbuscule formation. There is a decondensation of the plant's chromatin which indicates increased transcription of the plant's DNA in arbuscule containing cells [13]. Major modifications are required in the plant host cells to accommodate the arbuscules. The vacuoles shrink and other cellular organelles proliferate. The plant cell cytoskeleton is reorganized around the arbuscules [14].

The term Vesicular Arbuscular Mycorrhiza was originally applied to symbiotic associations formed by all fungi in the Glomales, but because a major suborder lacks the ability to form vesicles in roots, AM is now the preferred acronym. The AM type of symbiosis is very common as the fungi involved can colonize a vast taxonomic range of both herbaceous and woody plants, indicating a general lack of host specificity among this type. However, it is important to distinguish between specificity, innate ability to colonize, infectiveness, amount of colonization and effectiveness, plant response to colonization. AM fungi differ widely in the level of colonization they produce in a root system and in their impact on nutrient uptake and plant growth [15].

Biochemical Interactions between Host and Fungus

The biochemical interaction between the host and fungus seems to be very complex and influenced by number of interrelated biochemical, physiological and environmental processes. There occurs a reciprocal relationship between the fungus and the plants host but it is difficult to interpret the exact contribution that either organism leads to the association [16]. It is clear that the fungal symbiont must enter and maintain a parasitic relationship with its host for procurement of organic compounds required for its growth and reproduction.

Growth Regulating Substances

In pure culture, mycorrhizal fungi produces auxins, cytokinins, gibberellins and vitamins [17-19] and the effects of these compounds on plant growth and development are well documented [20]. The type and quantity of these substances produced by both partners could effect their association and overall response of the plants. Even though auxin appears to be the most common substance inducing mycorrhiza-like structures, others have found activity with colchicines, kinetin and various vitamins [19]. The development of short, swollen roots in some mycorrhizal infected plants may be due to the involvement of cytokinin in the symbiotic relationship. Mycorrhizal fungi synthesizes cytokinin in pure culture [17, 18]. However, all the mycorrhizal fungi are not capable of synthesizing cytokinins or of synthesizing them in sufficient quantities.

Carbohydrates

The fungus needs simple carbohydrates for its growth. Auxin influences the translocation of sugar from starch reservoirs of the plant [21] and also the hydrolysis of starch into sugar [22-24]. Since the mycorrhizal fungi generally assimilate soluble carbohydrates, their absence or presence in small quantities could influence mycorrhizal formation [25, 26]. The glucose derived from the host may get converted into fungal carbohydrates i.e. trehalose and mannitol which could not be metabolized by the host plant. The fungus intervenes in the carbohydrate metabolism of the host by acting as a sink [16]. In endomycorrhizae, the carbohydrate storage process may be different. Trehalose and mannitol have not been found in endomycorrhizal fungi [27-29].

Environmental Influences

The abovesaid carbohydrates and growth hormones or other substances are not alone responsible for mycorrhizal formation and maintenance of the symbiosis. Some environmental factors like light, soil conditions (moisture, mineral nutrients, pollutants) and also interaction with other soil organisms are also responsible for interactions between the host plant growth and development and mycorrhiza. Lower level of nitrogen (N) content leads to greater level of ectomycorrhizal formation [26, 29]. However, mycorrhizal development is not always indicative of nutritionally poor soils, but also may be dependent upon rooting behaviour of trees, soil types and balance of nutrients [16]. Besides this, the rate of fertilizer release and atmospheric pollutants may also influence the mycorrhizal development inhabiting an area. Endomycorrhizae is more prevalent in low fertility soils, especially when nitrogen and phosphorus are low [9]. Mycorrhizal infection is found to be best at high light intensities in few cases. Most of the mycorrhizal fungi need an optimum temperature for establishment of the symbiotic relationship and survival of the mycorrhizal conditions.

Role of Mycorrhiza in Crop Plants

Mycorrhiza plays a wide role in host plants. The fungus has greater applicability in enhancing plant growth under tough environmental conditions. The major roles by mycorrhizal association in the associated plants are briefly discussed below.

Nutrient Uptake by Plants

The mycorrhizal role in the mineral nutrition of plants is revealing in world-wide production of field and green house horticultural crops. The mycorrhizal association particularly with vegetables is found to improve the absorption of almost all the nutrients required by them for their growth such as phosphorus, copper, zinc, sulphur, magnesium, manganese, iron etc. The benefit of mycorrhizae to crop plants is mainly attributed to increased uptake of nutrients especially phosphorus. An increase in the carbon supplied by the plant to the Arbuscular Mycorrhizal fungi increases the uptake of phosphorus and the transfer of phosphorus from fungi to plant has been already proved. The increase in uptake of nutrients may be due to increased surface area of soil contact, increased movement of nutrients into mycorrhizae, modification in the root environment and increased storage [30]. The mycorrhizal associations with crops in nutrient deficient soils are found to absorb larger amount of nutrients than the non-mycorrhizal ones [8].

Solubilization of Plant Nutrients

More root exudates are secreted by the mycorrhizal roots system in the rhizosphere resulting in enhanced activity of useful rhizospheric microbes such as phosphate solubilizing microorganisms, organic matter decomposers and symbiotic as well as non-symbiotic biological nitrogen fixers etc. [8]. Mycorrhizae lower the rhizosphere pH due to selective uptake of ammonium ions and release of hydrogen ions. Decreased soil pH increases the solubility of phosphorus precipitates. The hyphal uptake of ammonium ions also increase the flow of nitrogen to the plant as NH_4^+ is adsorbed to the soil's inner surfaces and must be taken by diffusion [14].

Stress Tolerance

The mycorrhizal plants were able to produce better than non-mycorrhizal plants under a variety of stress situations. Many mycorrhizal fungi possess specific individual traits with respect to tolerance to soil temperature extremes, pH, moisture, low fertility, salinity, toxicants, etc., which may provide the host plant with an ecological competitive advantage facilitating increased plant survival, growth, nutrition and/or yield under stress conditions [31]. Moreover, by exploitation of larger soil volume, extended root growth and increased absorptive area, the mycorrhizal plants exhibit better growth than the non-mycorrhizal ones especially in the arid and semi-arid regions where low moisture and high temperature are very critical for survival and growth of plants. Mycorrhizal fungi are also involved in detoxification of soils high in metal toxicants. Heavy metal tolerance mechanisms are also known to be present in mycorrhizal fungus [32]. Endomycorrhizal fungi are known to have more impact on horticultural crop production and revegetation of adverse sites than ectomycorrhizal fungi because of their broader host range.

Utilization of Fixed Phosphates and Insoluble Phosphates

Recent advances on mycorrhizal research suggest that the symbiotic mycorrhizal association can lead to more economical use of phosphate fertilizers and better exploitation of cheaper and less soluble rock phosphates. The better utilization of sparingly soluble rock phosphate is explained by the hyphae making closer physical contact than the roots with the ions dissociating at the particle surface [8]. In roots, inorganic P appears to move at about 2 mm/hour [33]; in fungal hyphae the movement is 2 cm/hour [34] indicating a greater efficiency in uptake and possibly the presence of an active transport mechanism such as cytoplasmic streaming in the mycorrhizal fungus hyphae.

Disease Resistance

Mycorrhiza is found to offer adequate protection to the root system from the attack of pathogenic fungi. Wilt, root and stem diseases in horticultural crops cause large economic losses annually. These root-invading pathogens influence the soil type, pH, temperature, fertility, moisture and the presence of other soil-borne organisms. The mycorrhizal fungi such as *Lectarious delicious* and *Boletus* sp. antagonizing *Rhizoctonia solani*, *Lectarious camphorates*, *Lectarious* and *Cortinarious* sp. have been found to produce antibiotics known as 'chloromycorrhiza' and 'Mycorrhizine A' which are antifungal to the phytopathogens like *Rhizoctonia solani*, *Pythium debarynum* and *Fusarium oxysporum* etc. If the mycorrhizal fungus gains entry into root system prior to infection of root by the fungal pathogen, the fungal mantle in the mycorrhizal roots offers physical resistance to various soil-borne pathogenic fungi. Lignin production is also found to be enhanced in the mycorrhizal roots with the accumulation of some polysaccharides on the cell wall. In such tissues, the growth of pathogens like *Fusarium oxysporum* and *Pyrenochaeta terrestries* etc. have been found to be considerably restricted [8]. Such use of mycorrhiza is welcomed as a safe and environmentally acceptable alternative to the use of chemical measures for disease control. Various mechanisms exist including the use of microbial antagonists that produce antibiotics or lytic enzymes, that compete for nutrients with the pathogen, that directly invade and kill the pathogen as hyperparasites, that invade and transmit viral a virulence (hypovirulence) factors, or that are non-pathogenic but trigger or stimulate natural defense mechanisms in the host (induced resistance and cross protection). Another biological mechanism is the use of microbial agents to modify the chemical environment (allelopathic biocontrol) through the breakdown of organics in the soil to release antimicrobial compounds such as phenolics or to enhance the chelation of essential nutrients for the pathogen such as iron.

Phytoremediation

A significant long-term improvement in the soil quality parameters was noticed when the soil was inoculated with a mixture of indigenous Arbuscular

Mycorrhizal fungi species compared to the non-inoculated soil and soil inoculated with a single exotic species of Arbuscular Mycorrhizal fungi [35]. Various benefits like increase in plant growth, soil nitrogen content, higher soil organic matter content and soil aggregation were obtained. These might occur due to higher legume nodulation in the presence of Arbuscular Mycorrhizal fungi, better water infiltration and soil aeration due to soil aggregation.

Growth Response of Horticultural Crops to Mycorrhizal Fungi

Over millions of years, mycorrhizal fungi and plants have formed a mutual dependence. The fungi are nourished by root exudates and in return bring great amounts of soil nutrients and moisture to their host plants. Mycorrhizal plants can uptake 100 times or more nutrients than one without the beneficial fungi. The addition of mycorrhizal fungi spores to transplant roots, garden soils, potting soil, lawns or seeded crops will ensure the presence of these valuable plant allies. The difference in plant health and performance can be dramatic, especially when dealing with less-than-perfect soils. Some extremely dependent plants, including grapes, citrus, melons, oaks and pines may quire literally starve to death in soils that lack this helpful fungi [36].

The influence of AM fungi on the growth and nutrition is well documented in a number of horticultural crops. Most of the horticultural crops perform better and are more productive when well colonized by AM fungi. AM symbiosis increases the phosphorus and micronutrient uptake and growth of their plant host. Enhanced rooting of cuttings also was obtained with some mycorrhizal fungi. The nature of the root system apparently influences the response of a plant species to mycorrhizal fungi. A significant increase in plant growth and yield of several horticultural crops inoculated with AM fungi have been reported by several workers in crops like Easter lily [37], Poinsettia [38] tomato and capsicum [39], Chrysanthemum and China aster [40]. The yield of onion from *Glomus mosseae* inoculated seedlings was generally about twice those from uninoculated seedlings [41]. The improved growth performance and P-use efficiency by inoculating pre-germinated celery seeds with VAM fungus, *Glomus intraradices* [42].

The mycorrhizal inoculated citrus rootstocks become ready for budding 4-5 months early compared to uninoculated plants [43]. A similar trend was observed in cashew rootstocks with early rooting and faster establishment [44]. The extended vase life of chrysanthemum flowers besides the increase in flower yield was noted when inoculated with *Glomus mosseae* [40]. Troyer citrange rootstocks inoculated with *Glomus macrocarpum*, *G. caledonicum*, *G. velum*, *G. monosporum* and *Gigaspora margarita* became ready for budding 5-6 months earlier besides greatest improvements in growth and nutrition resulting in larger leaf area, plant height, stem diameter and plant biomass with higher P, Zn and

Cu contents [43]. The VAM fungus *Glomus intraradices* increased root and shoot growth of the citrus root stocks viz., Sour orange (*Citrus aurantium* L.) and 'Carriza' Citrange (*Citrus sinensis* (L.) OSb x *Poncirus trifoliata* (L.) Raf) [45]. Mycorrhizal colonization was also confirmed in kiwi and papaya trees [46]. Micropropagated plants are very delicate and should be hardened before transferring to the main field. During this hardening about 20-40% mortality is noticed. Inoculating the micropropagated plantlets with AM fungi after hardening found to improve plantlet vigour and growth in coffee, grapevine, apple, avocado, pineapple, kiwifruit, strawberry, raspberry, asparagus and banana [47-49].

In turmeric (*Curcuma longa*), combined application of FYM, azospirillum, phosphobacteria and VAM exhibited higher rhizome yield (33313.00 kg ha^{-1}) as compared to the application of 50 per cent recommended dose of fertilizer + phosphobacteria (18636.00 kg ha^{-1}). The highest curing percentage (21.09 per cent), cured rhizome yield and quality parameters like curcumin and oleoresin content were also regimed supreme by application of FYM, azospirillum, phosphobacteria and VAM [50-54].

Advantages of Mycorrhizal Bio-fertilizer Inoculation

It enhances seed germination, flowering and maturity with increased production. Its requirement is very small and also benefits the subsequent crop. Nutrient recycling is also enabled. It provides plant nutrients like phosphorus, potassium, calcium, magnesium, sulphur, iron, manganese, zinc and copper at minimum cost. It enhances plant growth by release of vitamins and hormones and plant growth substances like auxins and cytokinins etc. It helps in survival and proliferation of beneficial microorganisms like phosphorus solubilizers, organic matter decomposers and nitrogen fixing etc. It increases the availability and uptake of secondary and micronutrients which are in relatively insoluble and immobile nature. It improves the physical, chemical and biological properties of soil by organic matter decomposition and soil aggregation. The crop yield is increased to about 20 to 40%. It is an eco-friendly approach enhancing the soil fertility and plant growth. It saves the phosphate requirement at about 20-40 kg/ha. Secretion of antibiotics competes or antagonizes pathogenic microorganisms aiding in disease suppression. It leads to better adaptation of plant to adverse environment conditions like drought by modifying soil-plant-water relations. It increases establishment, nodulation and atmospheric nitrogen fixation capacity of leguminous crops.

Conclusion

Now, it is very clear that Vesicular-Arbuscular Mycorrhizal fungi can enhance the growth and production of horticultural crops. Though the potential benefits of mycorrhizal fungi have been discussed, there is only limited information on the role of mycorrhizal fungi in horticultural crop production. It is important to determine the mycorrhizal dependency of the horticultural crops grown in a region and select the mycorrhizal dependent plants for application of inoculation techniques. Further researches are required for understanding the physiology and ecology of the association and host specificity i.e. particular cultivar benefited from a particular strain of mycorrhiza. The factors deleterious to mycorrhizal like pesticides, high level of inorganic fertilizers, fumigation and poor aeration etc. should be further investigated. Hence, with the abovesaid research advancements, the mycorrhizal fungi can be utilized effectively in horticultural crops with reduced cost of cultivation besides creating sustainable environment.

References

1. Harrison, M.J. *Annu Rev Microbiol.* 2005, 59:19-42.
2. Selosse, M.A., Richard, F., He, X. et al. *Trends Ecol Evol.* 2006, 21:621-628.
3. Li, H., Smith, S.E., Holloway, R.E. et al. *New Phytol.* 2006, 172:536-543.
4. Wang, B. and Qiu, Y.L. *Mycorrhizahello* 2006, 16(5):299-363.
5. Trappe, J.M. *Bot. Rev.* 1962, 28:538-605.
6. Meyer, F.H. *In:* Ectomycorrhizae. Marx, G.C. and Kozlowski, T.T. (eds), Academic Press, New York, 1973, pp. 79-105.
7. Maronek, D.M., Hendrix, J.W. and Kiernan, J. *In:* Horticultural Reviews. Janick. J. (ed.), AVI Publishing Company, Connecticut, 1981, 3:172-213.
8. Mishra, B.B. and Mishra, S.N. *Orissa Review.* http://orissagov.nic.in/e-magazine/ Orissareview/dec2004/englishPdf/mycorrhizaanditssignificanceinsustainable-forest.pdf, 2004.
9. Gerdemann, J.W. and Trappe, J.M. *In:* Endomycorrhizas. Sanders, F.E., Mosse, B. and Tinker, P.B. (eds), Academic Press, New York, 1975:35-51.
10. Gerdemann, J.W. and Nicholson, T.H. *Trans. Brit. Mycol. Soc.* 1963, 46:235-244.
11. Muchovej, R.M. 2001, http://edis.ifas.ufl.edu/pdffiles/AG/AG11600.pdf
12. Mikola, P. *Acta. For. Fenn.* 1965, 79:1-56.
13. Gianinazzi-Pearson, V. *The Plant Cell* 1996, 8:1871-1883.
14. http://en.wikipedia.org/wiki/Arbuscular_mycorrhiza
15. http://cropsoil.psu.edu/sylvia/mycorrhiza.htm
16. Meyer, F.H. *Annu. Rev. Plant Physiol.* 1974, 25:567-586.
17. Miller, C.O. *In:* Mycorrhizae. Hacskaylo, E. (ed.), GPO, Washington, DC, 1971, pp. 168-174.
18. Crafts, C.B. and Miller, C.O. *Plant Physiol.* 1974, 54:586-588.
19. Slankis, V. *In:* Ectomycorrhizae. Marx, G.C. and Kozlowski, T.T. (eds), Academic Press, New York, 1975, pp. 231-298.
20. Torrey, J.A. *Annual Rev. Plant Physiol.* 1976, 27:435-459.
21. Thimann, K.V. *In:* Plant physiology. Steward, F.C. (ed.), Academic Press, New York, 1972, p. 365.
22. Borthwick, H.A., Hammer, K.C. and Parker, M.W. *Bot. Gaz.* 1937, 98:491-519.

23. Alexander, T.R. *Plant Physiol.* 1938, 13:845-858.
24. Bausor, S.S. *Bot. Gaz.* 1942, 104:115-121.
25. Bjorkman, E. *Stud. For. Suec.* 1970, 83:1-24.
26. Bjorkman, E. *Symb. Bot. Upsal* 1942, 6:1-191.
27. Hepper, C.M. and Mosse, B. *Trehalose and mannitol in vesicular-arbuscular mycorrhizae*, Rothamsted. Expt. Sta., Harpenden, Herts, UK, 1972.
28. Bevege, D.I., Bowen, G.D. and Skinner, M.F. *In:* Endomycorrhizas. Sanders, F.E., Mosse, B. and Tinker, P.B. (eds), Academic Press, New York, 1975, pp. 149-174.
29. Marx, D.H., Hatch, A.B. and Mendicino, J.F. *Can. J. Bot.* 1977, 55:1569-1574.
30. Bolan, N.S. *Plant and Soil* 1991, 134:189-207.
31. Trappe, J.M. *Annu. Rev. Phytopathol.* 1977, 15:203-222.
32. Ashida, J. *Annu. Rev. Phytopathol.* 1965, 3:153-174.
33. Crosset, R.N. and Loughman, B.C. *New Phytol.* 1966, 65:459-468.
34. Littlefield, L.J. *Physiol. Plant.* 1966, 19:264-270.
35. Jeffries, P., Gianinazzi, S., Perotto, S., Turnau, K. and Barea, J. *Biology and Fertility of Soils* 2003, 37:1-16.
36. http://www.bio-organics.com/
37. Ames, R.C. and Linderman, R.G. *Can J. Bot.* 1978, 56:2773-2780.
38. Barrows, J.B. and Rancadori, R.W. *Mycologia* 1977, 69:1173-1184.
39. Mallesha, B.C. and Bagyaraj, D.J. *J. Soil Biol. Ecol.* 1997, 17:34-38.
40. Mamatha, K.B. and Bagyaraj, D.J. *Proc. Nat. Acad. Sci. India* 2001, 71(B):157-163.
41. Snellgrove, R.C. and Stribley, D.P. *Plant and Soil* 1986, 92(3):387-397.
42. Ronald, W.R., Lawrence, E.D., Richard, N.R. et al. *Journal of plant nutrition* 2002, 25(8):1839-1853.
43. Vinayak, K. and Bagyaraj, D.J. *Biology and Fertility of Soils* 1990, 9(4):311-314.
44. Lakshmipathy, R., Balakrishna, A.N., Bagyaraj, D.J. et al. *The Cashew* 2000, 14:20-24.
45. Dutra, P.V., Abad, M., Almela, V. et al. *Scientia Horticulturae* 1996, 66(1-2):77-83.
46. Schubert, A. and Cravero, M.C. *Acta Hort.* 1990, 282:199-202.
47. Varma, A. and Scheup, H. *In:* Mycorrhizae: Biofertilizers for the future. Adholeya, A. and Singh, S. (eds), Tata Energy Research Institute, New Delhi, 1995, pp. 322-328.
48. Shashikala, B.N., Reddy, B.J.D. and Bagyaraj, D.J. *Crop Res.* 1999, 18:900.
49. Mathews, D. and Hedge, R.V. and Sreenivasa, M.N. *Karnataka J. Agric. Sci* 2003, 16:438-442.
50. Velmurugan, M., Chezhiyan, N. and Jawaharlal, M. *Asian Journal of Soil Science* 2007, 2(2):113-117.
51. Velmurugan, M., Chezhiyan, N. and Jawaharlal, M. *Asian Journal of Horticulture* 2007, 2(2):23-29.
52. Velmurugan, M., Chezhiyan, N., Jawaharlal, M. et al. *Asian Journal of Horticulture* 2007, 2(2):291-293.
53. Velmurugan, M. and Chezhiyan, N. *South Indian Horticulture* 2007. 53(1-6):392-396.
54. Velmurugan, M., Davamani, V., Anand, M. et al. *Plant Archives* 2008, 8(1):9-32.

10

Mycobization as a Biotechnological Tool: A Challenge

M.S. Velázquez and M.N. Cabello

Instituto de Botánica Spegazzini, Facultad de Ciencias Naturalesy
Museo, Universidad Nacional de La Plata, La Plata
Provincia de Buenos Aires, Argentina

Introduction

At the beginning of the twentieth century, agricultural products provided food to a world population of about 1200 million people, whereas nowadays they do for a population of about 6000 million. Due to the increasing population growth rate which is about 60 million people/year by 2020 the population will be around 8000 million. Some projections indicate that by then the world production of the main crops will have to be increased by about 700 million tons. The speed of these changes leads to the need to pay attention to the consequences brought about by this demand, such as the massive use of agrochemicals, which, in turn do not allow the maintenance of soil productivity.

The stability and productivity of an agroecosystem or of a natural system depend, to a large extent, on the quality of the soil, not only to obtain profitable crops, which guarantee the sustainability of the agroecosystem, but also to minimize the environmental impact [1]. In the past, much attention was paid to investigate the physical and chemical components of the soil [1], whereas little attention was paid to its biological properties. However, in the last years, more studies have emphasized the importance of keeping a diverse and active microbial community [2, 3], in order both to guarantee the quality of the soil and to obtain greater amounts of economical and healthy foods [4].

Microorganisms populations are involved in a net of interactions that affect soil-plant development, and some of the beneficial activities of the microorganisms can be exploited as low-cost biotechnological tools to solve sustainability problems [2]. The development reached by means of these biotechnological techniques has opened new possibilities to produce a wide range of different products. Currently, there are applications related with human health, feeding and agriculture, and industrial and mining processes, which take care and preserve the environment.

Among beneficial soil organisms, we can mention arbuscular mycorrhiza fungi that belong to the phylum Glomeromycota [5], which form a universal symbiosis established with more than 80% of plant species, including most agricultural and horticultural crops, and herbaceous and arbustive species in natural ecosystems [6].

Arbuscular mycorrhizal fungi are key components in all ecosystems, not only because of the biomass they provide to the soil, but also because they are closely related to processes that take place in the rhizosphere, establishing synergic relationships with other microorganisms that benefit the plant. They act actively in the uptake and translocation of nutrients, especially of phosphorus, benefiting both plant growth and production. In addition, they provide greater tolerance to pathogens, thus contributing to the plant's health. They decrease the incidence of diseases due to changes in the immunological system and to a better nutritional state [7]. As a result, mycorrhized plants frequently exhibit an increase in the growth, production and survival as compared to non-mycorrhizal ones.

Considering all the beneficial aspects mentioned, arbuscular mycorrhizal fungi are potentially useful as inoculants to achieve a sustainable production, and substantially decrease the need for agrochemicals and guarantee the ecological equilibrium. At a commercial scale, important benefits can be derived from their use, including a better plant growth and greater production, an improvement in the uniformity of the crop, a reduction in phosphorous fertilization, a decrease in root damage, and a reduction in harvest time and environmental stress.

Potential Host Plants

As previously mentioned, arbuscular mycorrhizal fungi have an extremely wide range of host plants, although not all plants present the same degree of mycorrhizal dependence. Janos [8] defined mycorrhizal dependence as 'the inability of plants without mycorrhizas to grow or survive without some increase in soil fertility'.

Mycotrophic plants differ in their colonization percentages. Some crops, such as sweet potato, soybean, maize, sorghum, barley, sugarcane, tobacco, cotton and cacao, frequently exhibit high colonization rates under natural conditions. Other crops such as wheat, beans, coffee and tomato can have more moderate colonization rates. In addition, some small intraspecific differences can be observed in the colonization percentages between different ecotypes, cultivars or clones of the same crop. Additionally, the environmental conditions also affect the formation of arbuscular mycorrhizas and/or the root extent.

For instance, crops with high supplies of fertilizers rich in nitrogen, phosphorus and potassium, inhibit mycorrhizal development.

Although most agricultural crops are potential host plants for arbuscular mycorrhizal fungi, their use as biofertilizers is limited by several reasons: one of them is that most agricultural soils present a varied and abundant mycorrhizal biota, generally adapted to the crop's condition, which can compete until eliminating the biological inoculant added. Due to the large extensions of cultivated areas, in order to guarantee a successful colonization, it is necessary to apply a great amount of inoculant to prevent its dilution; besides, as previously mentioned, it should be considered that if the system has important supplies of phosphorus, this will either decrease or inhibit the colonization.

Instead, their use is mainly restricted to horticultural crops and to diverse plants that either propagate by stolons or micropropagate—aromatic and ornamental plants—and share one characteristic: they go through a pot period where the inoculation strategies are simpler, environmental variables can be controlled, and a starter substrate completely free of microorganisms can be selected in order to guarantee a successful colonization.

The Use of Phosphorus in the Soil

Phosphorus is the most important plant growth-limiting nutrient in soil besides nitrogen. It is estimated that 5.7 billion hectares in the world do not contain the necessary amounts of available phosphorus to sustain the productivity of their crops [9]. Phosphates in the soil are present under three forms: (1) soluble inorganic phosphorus in the soil solution, (2) insoluble inorganic phosphorus in crystalline nets, and (3) phosphorated organic compounds such as phytate [10]. They are relatively immobile and diffuse slowly to the plant roots [11] and capture phosphorus at a greater speed than that at which the nutrient diffuses. In soils with a low availability of phosphorus, this generates areas of depletion around the roots [12], which is a critical problem for the uptake and cycling of nutrients in soil-plant systems [5]. Here is where the mycorrhizal activity is centered, since the external hyphae can explore a higher volume of the soil, overcoming the depletion areas (Figure 1).

It is important to point out that both mycorrhized and non-mycorrhized roots absorb phosphorus from the same pool available in the soil [13-15]. Mycorrhizal fungi are not able to solubilize the non-available forms of phosphorus, but can increase its absorption by providing them a greater exploration of the soil through the external mycelium. On the other hand, mycorrhizal hyphae have a higher affinity for phosphate, as expressed in the Michaelis-Menten equation by a lower Km value and absorb phosphorus at lower solution concentrations than roots do [16].

As regards phosphate cycling, two general types of microbiological processes have been described: (1) those that promote the solubilization of non-soluble phosphorus sources in the soil, and (2) those that increase the

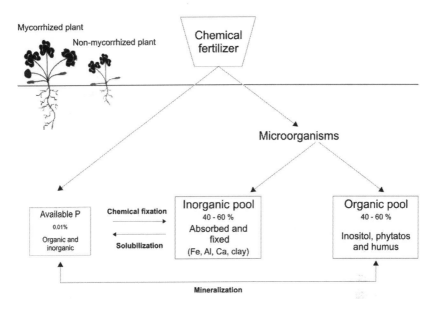

Figure 1: Phosphorus cycle in the soil; the values are estimated percentages in the mineral soil.

uptake of soluble phosphates around the plant. The former is carried out by solubilizer microorganisms, among which there is a great number of phosphate-solubilizing bacteria [17], actinomycetes [18] and fungi [19]. These organisms are capable of solubilizing the phosphate content in minerals by means of secretion of carboxylic acids, enzymes and bacterial mucilage [19-21]. The second mechanism is carried out by means of mobilizing microorganisms, which are basically arbuscular mycorrhiza-forming fungi [7].

Interactions in the Mycorrhizosphere

In the soil, the populations of microorganisms concentrate around the roots, stimulated by root excretions and exudates. In this area defined as the rhizosphere [22-25], there is an increase in the activity of microorganisms such as bacteria, saprotrophic fungi, pathogens, and symbionts, which are involved in a wide range of important functions involved in plant growth. The development of the rhizosphere is a dynamic process that incorporates physical, chemical and biological modifications that occur in the root-soil interphase. An important consequence of the increase in microbial activity of the rhizosphere derives from the ability of microorganisms to alter the availability of nutrients for the plant.

The interactions that occur at the soil-root interphase are essential to understand the dynamics of the processes that are characteristic of the rhizosphere [26, 27]. A great number of microorganisms of the rhizosphere develop activities that alter the availability of nutrients and act synergistically with mycorrhizal fungi to increase plant growth. These interactions are critical because they regulate the formation and development of mycorrhizas, which, in turn, can affect the microbial population of the rhizosphere. This deserves special attention for the development of biotechnological techniques that make use of the microorganisms of the rhizosphere. Below we will detail some of the most relevant interactions between arbuscular mycorrhizal fungi and microorganisms of the rhizosphere.

Saprotrophic Fungi

Saprotrophic fungi from the soil that inhabit in the rhizosphere interact with arbuscular mycorrhizal fungi in different ways, and can have either positive, negative or neutral effects in the establishment and later development of the symbiosis. Understanding this synergism is extremely important because mycobization—co-inoculation of selected strains of saprotrophic and mycorrhizal fungi—in agriculturally important host plants might favour faster root colonization and thus benefit plant nutrition. On the other hand, the selection of saprotrophic fungi with inhibitory effects on the mycorrhizal symbiosis may allow a greater control on the quality and purity of a mycorrhizal inoculum.

Numerous investigations have evaluated the interaction between mycorrhizal and saprotrophic fungi. However, there are only few studies about the synergistic effect between yeast and arbuscular mycorrhizas though they are the two important components in the microbial communities of the soil. Singh et al. [28] studied the effects of inoculation with the commercial yeast *Saccharomyces cerevisiae* Hensen in leguminous plants (*Leucaena leucocephala, Glycine max, Cajanus cajan, Phaseolus mungo, P. aureus* and *Vigna unguiculata*) growing on a substrate with native arbuscular mycorrhizal fungi. They evidenced a significant increase in plant biomass, as well as an increase in the parameters of the mycorrhizal colonization, in the number of spores and the formation of arbuscules and vesicles in the presence of *S. cerevisiae*. Fracchia et al. [29], on the other hand, found that the yeast *Rhodotorula mucilaginosa* (Jörg) Harrison stimulated hyphal growth of *Glomus mosseae* (Nicol. & Gerd.) Gerd. & Trappe and *Gigaspora rosea* Nicol. & Schenk and that the exudates produced by *R. mucilaginosa* increase the colonization of soybean by *G. mosseae* and of red clover by *G. rosea*.

Fracchia et al. [30] also investigated different combinations between the saprotrophic *Fusarium oxysporum* Schlecht. Fr. and several species of arbuscular mycorrhizal fungi such as *G. mosseae, G. fasciculatum* Gerd. & Trappe, *G. intraradices* Schenk & Smith, *G. clarum* Nicol. & Schenk and *G. deserticota* Trappe, Bloss & Menge on agricultural (maize, sorghum and wheat) and horticultural (lettuce, tomato, lentil and pea) crops. They found that *F. oxysporum* influences the mycorrhizal colonization and the growth response in host plants, but that this capacity varies depending on the plant and mycorrhizal fungus studied. These results suggest a direct interaction between saprotrophic and mycorrhizal fungi.

Phosphorus Solubilizers

In the soil, there are also different microorganisms and other fungi capable of solubilizing phosphates, as well as of increasing the absorption of phosphorus in the plant [31]. Velázquez et al. [32] have proposed the term 'Mycobization' to denominate the inoculation of plants with different functional groups belonging to the Kingdom Fungi, as is the case of the co-inoculation with solubilizing saprotrophs and mobilizing symbiotic biotrophs.

The combined inoculation of arbuscular mycorrhizal fungi and phosphorus-solubilizing microorganisms has allowed a better absorption of native phosphorus and of the phosphorus in the rock phosphate [33, 34]. Rock phosphate can be a fertilizer of effective agronomic use and low cost mainly in highly meteorized acid soils, since it provides phosphorus and calcium, two elements that are generally lacking in these soils [35-37]. Numerous studies involve the use of rock phosphate as a source of phosphorus. For instance, Graham and Timmer [38, 39] have evaluated the interaction of the arbuscular mycorrhizal fungus *G. intraradices* in different citrus trees (citrange, lemon rouge and orange sour), with different levels of superphosphate and rock phosphate. They demonstrated a more significant growth of the plants fertilized with rock phosphate, being this form of P more permanent and available for a longer time than the P supplied in the superphosphate.

On the other hand, Omar [40] evaluated the capacity of solubilizing P from rock phosphate of the geofungi *Aspergillus niger* van Tieghem and *Penicillum citrinum* Thom, in wheat plants inoculated with the mycorrhizal fungus *Glomus constrictum* Trappe. Positive effects were recorded in plant growth and content of phosphorus in those plants co-inoculated with *G. constrictum* and *A. niger* or *P. citrinum* and in those in which rock phosphate was used as the phosphorus source. In addition, increases have been recorded in the growth of wheat plants growing in a sandy substrate with a deficit in nutrients when inoculated with two phosphoric rock-solubilizing microorganisms—*Bacillus circulans* Jordan and *Cladosporium herbarum* Link: Fries—with a mycorrhizal fungus belonging to the *Glomus* genus [41].

The combination of rock phosphate, arbuscular mycorrhizal fungi and phosphorus-solubilizing fungi presents an interesting alternative to the use of fertilizers worldwide [32, 33, 42]. Due to the importance of microbial solubilization and phosphorus mobilization processes, Cabello et al. [33] evaluated the effects of the co-inoculation of *G. mosseae* (a symbiont biotroph) and *P. thomii* Maire (a solubilizing saprotrophic fungus) in the growth of *Mentha piperita* L. plants and the percentage of root mycorrhizal colonization along two harvest times. This experiment involved a series of five treatments on perlite:vermiculite (inert substrate) with different fungal combinations, rock phosphate and non-inoculated controls. It was found that *P. thomii* neither affected nor stimulated the capacity of *G. mosseae* for colonizing roots (Table 1). The maximum growth in both non-inoculated plants and plants inoculated with *G. mosseae* and rock phosphate was observed after the first harvest (Table 2). These results are in agreement with those found by Graham and Timmer [38], where the maximum values of growth were recorded at the moment in which phosphorus was added to the medium. The fact that biomass increased in mycorrhized plants after the first harvest represents important economic and agroeconomic benefits, since *M. piperita* is a perennial plant that can be harvested up to four times along its life cycle.

On the other hand, the phosphorus present in the plant tissues was greater after the first harvest in all the treatments. It was also observed that phosphorus was solubilized in both the mycorrhized and the non-mycorrhized root systems without the solubilizing fungus. This confirms that the abiotic factors can solubilize part of the P present in the system [43]. The roots can also contribute to the mobility and bioavailability of the phosphorus in the rhizosphere by means of the induction of areas of depletion/accumulation of phosphorus, acidification/alkalinization and exudation of organic acids/anions [9].

Table 1. Percentage of root colonization, hyphae, arbuscules and vesicules

Features	R.P + Gm	R.P + Gm + Pt
Colonization	94a	95a
Hyphae	27a	23a
Arbuscules	31a	64[a]
Vesicules	14.6a	4b

Gm, *Glomus mosseae*; Pt, *Penicillum thomii*; RP, rock phosphate; and Values are means ± SD for three pots. The same letter above the bar indicates that values did not differ significantly between treatments according to ANOVA and LSD test (P< 0.05) (Adapted from [33]).

Table 2. Dry weight in plant and P content in substrate

Features	p:v		p:v+RP		p:v+ RP +Gm		p:v+ RP +Pt		p:v+RP+Gm+Pt	
Harvest	I°	II°	I°	II°	I°	II°	I°	II°	I°	II°
Dry weight (g/plant)	2.14a	1.90a	5.84c	2.45[a]	4.42ab	6.54b	3.66b	4.72ab	4.59bc	4.74ab
P substrate (mg/kg substrate)	3.9a	3.75a	12.2a	16.36b	11.4a	12.74ab	37.03c	18.26c	64.83b	40.19b

I°: first harvest; II°: second harvest; p:v (perlite:vermiculita); RP (rock phosphate); Gm (*Glomus mosseae*) and Pt (*Penicillum thomii*); and Values are means + SD for three pots. The same letter above the bar indicates that values did not differ significantly (Adapted from [33]).

Velázquez et al. [32] combined the use of rock phosphate as a source of phosphorus, *G. mosseae* as a mobilizing microorganism, and *Aspergillus niger* as a solubilizer, in *Lycopersicon esculentum* Mill plants. The experiments involved different treatments developed on a perlite:vermiculite:sand substrate (inert substrate) with different fungal combinations, rock phosphate and non-inoculated controls. They found that the number of *G. mosseae* spores was significantly different between treatments: the treatment inoculated with *G. mosseae* and without rock phosphate presented a higher number of spores, whereas the treatment inoculated with *G. mosseae* and rock phosphate evidenced the lowest value in the number of spores (Table 3). This result correlates with the nutritional deficit of the host plant, also reflected in the growth parameters evaluated and the low content of phosphorus in the aerial biomass. In conditions of nutritional stress, the flow of carbohydrates from the plant to the fungus decreases, and, as a consequence, sporulation is stimulated. Mycorrhizal colonization was positive in all the treatments inoculated with *G. mosseae*, although the colonization rates decreased when the content of P in the soil increased.

This negative correlation between the availability of phosphorus and the mycorrhizal colonization has been well documented [7]. A greater aerial biomass in the co-inoculated treatments, either with or without the application of rock phosphate has also been recorded (Figure 2). The greater values of phosphorus concentrations in the substrate were obtained in the treatment where rock phosphate and the solubilizer *A. niger* was added, thus confirming the solubilizing capacity of this microorganism.

This work confirmed the synergistic effects between mycorrhiza-forming fungi and solubilizing microorganisms. It is also important to point out that the solubilizing capacity of the microorganisms tested *in vitro* does not always reflect the potential solubilizing capacity of these microorganisms in field conditions [19, 44]. In both *M. piperita-G. mosseae-P. thomii* and *L. esculentum-*

Table 3. Number of spores in 100 g dry substrate and percentage of root colonization

Treatment	No. of spores/100 g dry substrate	Percentage root colonized
Control	-	-
Gm	1496a	40ab
An + Gm	738b	50a
RP + Gm	421b	41ab
RP + An	-	-
RP + An + Gm	598b	32b

Gm, *Glomus mosseae*; An, *Aspergillus niger*; RP, rock phosphate; Treatment number between brackets; and Values are means ± SD for five pots. The same letter above the bar indicates that values did not differ significantly between treatments according to ANOVA and LSD test (P< 0.05) (Adapted from [32]).

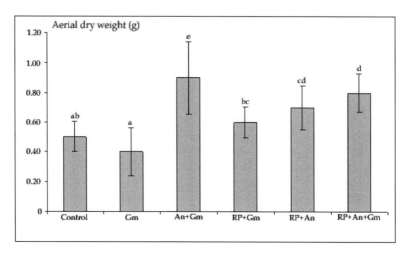

Figure 2: Aerial dry weight in tomato plants. Gm, *Glomus mosseae*; An, *Aspergillus niger*; and RP, rock phosphate. Between brackets are treatment numbers; Values are means ± SD for five pots. The same letter above the bar indicates that values did not differ significantly between treatments according to ANOVA and LSD test (P< 0.05). (Adapted from [32]).

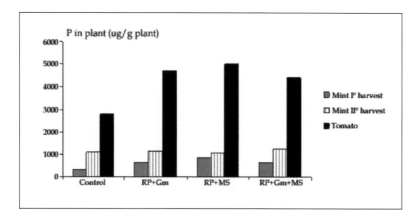

Figure 3: P content in *M. piperita* and *L. esculentum* plants. RP (rock phosphate), Gm (*G. mosseae*), and MS (solubilizing microorganism) (Adapted from [32, 33]).

G. mosseae-A. niger systems, the synergism between P-mobilizing and P-solubilizing microorganisms allowed the increment of P content in both plants (Figure 3). It can be observed that co-inoculation of *G. mosseae-A. niger* was more effective in increasing P contents in the plant than the co-inoculation with *G. mosseae-P. thomii*, for which it is important to point out that each soil-plant-fungus system requires special attention.

Conclusion

Considering that P is an essential nutrient for plant growth and taking into consideration that most soils are deficient in assimilable phosphorus, this review highlights the most important achievements related with the interactions between rhizospheric, P-solubilizing and P-mobilizing microorganisms and their potential as biofertilizers. Mycobization is a low-cost technological tool that does not have a negative impact on systems of intensive use such as organic orchards. However, further studies are necessary to evaluate the effects of mycobization in large-scale extensive crops in soils with a nutritional deficit under different agroclimatic conditions. In the first stage, we should evaluate the potentiality of the native microbiota in each environment in order to manage it by means of appropriate agronomical practices, without introducing foreign microorganisms, and thus be able to obtain a sustainable production and guarantee either a low or no environmental impact.

Acknowledgements

This research work was supported in part by grants from the Comisión de Investigaciones Científicas (CIC) de la Provincia de Buenos Aires and Universidad Nacional La Plata (UNLP). M.S. Velázquez is a recipient of scholarships from CONICET and M.N. Cabello is a researcher of the CIC.

References

1. Parr, J.F., Papendick, R.I., Hornick, S.B. et al. *American Journal of Alternative Agriculture* 1992, 7:2-3.
2. Kennedy, A.C. and Smith, K.L. *Plant Soil* 1995, 170:75-86.
3. Barea, J.M. *In:* Plant growth-promoting rhizobacteria, present status and future prospects. Ogoshi, A., Kobayashi, L., Homma, Y. et al. (eds), OCDE, Paris, 1997, pp. 150-158.
4. Dodd, J. *Outlook on Agriculture* 2000, 29:63-70.
5. Schüβler, A., Schwarzott, D. and Walker, C. *Mycological Research* 2001, 105:1413-1421.
6. Barea, J.M., Azcon, R. and Azcon-Aguilar, C. *In:* Microorganisms in Soils: Roles in Genesis and Functions. Buscot, F. and Varma, A. (eds), Springer-Verlag, Heidelberg, Germany, 2005, pp. 196-212.
7. Smith, S.E. and Read, D.J. Mycorrhizal Simbiosis. Academia Press London, UK, 1997.
8. Janos, D.P. *Mycorrhiza* 2007, 17:75-91.
9. Hinsinger, P. *Plant and Soil* 2001, 237:173-195.
10. Tinker, P.B.H. *In:* Endomycorrhizas. Sanders, F.E., Mosse, B. and Tinker, P.B. (eds), Academic Press, London, 1975.

11. Nye, P.H. and Tinker, P.B. Solute movement in the Soil-Water System. Blackwell Scientific, Oxford, 1977.
12. Jeffries, P. and Barea, J.M. *In:* Fungal associations. Hock, B. (ed.), Springer, Heidelberg, 2001, pp. 95-113.
13. Mosse, B. *New Phytologist* 1973, 72:127.
14. Nelson, C.E., Bolgiano, N.C., Furutani, S.C. et al. *Journal of the American Society for Horticultural Science* 1981, 106:786.
15. Joner, E.J. and Jakobsen, I. *Plant and Soil* 1995, 172:221-227.
16. Lange Ness, R.L. and Vlek, P.L.G. *Soil Science Society of American Journal* 2000, 64:949-955.
17. Taha, S.M., Mahmoud, S.A.Z., Halim El-Damaty, A. et al. *Plant and Soil* 1969, 31:149-160.
18. Rao, A.V., Venkatestwarlu and Kaul, P. *Current Science* 1982, 51:1117-1118.
19. Whitelaw, M.A. *Advances in Agronomy* 2000, 69:100-153.
20. Deubel, A. and Merbach, W. *In:* Microorganisms in Soils. Buscot, F. and Varma, A. (eds), Springer-Verlag, Heidelberg, Germany, 2005, pp. 177-191.
21. Kucey, R.M.N. *Applied and Environmental Microbiology* 1987, 53:2699-2703.
22. Bowen, G.D. *In:* Contemporary Microbial Ecology. Ellwood, D.C., Latham, M.J., Hedger, J.N. et al. (eds), Academic Press, London, 1980, pp. 283-304.
23. Foster, R.C. and Bowen, G.D. *In:* Phytopathogenic Prokaryotes. Mount, R. and Lacey, C. (eds), Academic Press, New York, 1982, pp. 159-185.
24. Lynch, J.M. Soil Biotechnology. Microbiological Factors in Crop Productivity. Blackwell Scientific Publications, Oxford, 1983.
25. Curl, E.A. and Truelove, B. The Rhizosphere. Springer-Verlag, Berlín, 1986.
26. Newman, E.I. *Biological Reviews* 1978, 53:511-554.
27. Suslow, T.V. *In:* Phytopathogenic Prokaryotes. Mount, R. and Lacey, C. (eds), Academic Press, New York, 1982, pp. 187-223.
28. Singh, C.S., Kapoor, A. and Wange, S.S. *Plant and Soil* 1991, 131:129-133.
29. Fracchia, S., Godeas, A., Scervino, M. et al. *Soil Biology and Biochemestry* 2003, 35:701-707.
30. Fracchia, S., Garcia-Romera, I., Godeas, A. et al. *Plant and Soil* 2000, 223:175-184.
31. Kucey, R.M.N., Janzea, H.H. and Leggett, M.E. *Advences in Agronomy* 1989, 42:199-228.
32. Velásquez, M.S., Elíades, L.A., Irrazabal, G.B. et al. *Journal of Agricultural Technology* 2005, 1:315-326.
33. Cabello, M., Irrazabal, G., Bucsinszky, A.M. et al. *Journal of Basic Microbiology* 2005, 45:182-189.
34. Goenadi, D.H. and Siswanto, Sugiarto, Y. *Soil Science Society of American Journal* 2000, 64:927-932.
35. He, Z.L., Baligar, B.C., Martens, D.C. et al. *Soil Science Society of American Journal* 1996, 60:589-595.
36. He, Z.L., Baligar, B.C., Martens, D.C. et al. *Soil Science Society of American Journal* 1996, 60:1596-1601.
37. Wright, R.J., Baligar, V.C. and Belesky, D.P. *Soil Science* 1992, 153:25-36.
38. Graham, J.H. and Timmer, L.W. *Journal of the American Society for Horticultural Science* 1984, 109:118-121.

39. Graham, J.H. and Timmer, L.W. *Journal of the American Society for Horticultural Science* 1985, 110:489-492.
40. Omar, S.A. *World Journal of Microbiology and Biotechnology* 1998, 14:211-218.
41. Singh, S. and Kapoor, K.K. *Biology and Fertility of Soils* 1999, 28:139-144.
42. Antunes, V. and Cardoso, E.J.B.N. *Plant and Soil* 1991, 131:11-19.
43. He, Z.L., Baligar, V.C., Martens, D.C. et al. *Plant and Soil* 1999, 170:131-140.
44. Villegas, J. and Fortín, J.A. *Canadian Journal of Botany* 2002, 80:571-576.

11

Phytoremediation: Biotechnological Procedures Involving Plants and Arbuscular Mycorrhizal Fungi

S.E. Hassan, M. St-Arnaud, M. Labreque and M. Hijri

Institut de Recherche en Biologie Végétale, Département de Sciences Biologiques, Université de Montréal, 4101 Rue Sherbrooke Est Montréal (Québec), H1X 2B2, Canada

Introduction

Environmental contamination is a serious issue originating from variable sources and applications, such as the use of agrochemical fertilizers, sewage sludge, and pesticides, or industrial activities, in particular metal mining, smelting, oil and gas operations, and other human uses [1, 2]. These applications release huge quantities of hazardous pollutants including organic and inorganic compounds into the air, water, and soil biospheres [3]. Inorganic contaminants involving heavy metals are natural components into the Earth's crust [4, 5]. Heavy metals freeing from various sources represent approximately 22,000 t of Cd, 9,39,000 t of Cu, 7,83,000 t of Pb, and 13,50,000 t of Zn through the last decades [6]. Heavy metals are classified as a group of 53 elements having a specific density higher than 5 g/cm^3 [7, 8]. Heavy metals can occur in soil for long periods of time, adhering to soil granules or polluting the underground water [9]. Among heavy metals, Cu, Fe, Mn, Ni, and Zn are essential elements required for normal plant growth [8], and these compounds have fundamental functions in nucleic acids metabolism; moreover, they are important for electron transfer, enzymatic catalyzation, as well as redox reactions [8]. On the contrary, other heavy metals like Cd, Pb, Hg, and As (as metalloids) are not required for living organisms [10].

Plant roots can uptake essential heavy metals from soil through specific and non-specific transporters [8], while non-essential elements are taken up by passive diffusion and wide specificity-metal transporters [11]. Huge quantities of heavy metals and metalloids (such as Cd, Pb, Zn, Cu, and As) hamper the biological activities for both higher organisms and soil microbes [12], and thereby heavy metals disturb ecosystem consistently involving microbial activities [13, 14]. High concentration of heavy metals shifts enzymatic functions by changing protein structure and replacing necessary components causing deficiency features [8]. Also, plasma membrane and its permeability are highly sensitive to heavy metal poisoning; hence, membrane functions have been influenced by modification of the role of membrane's protein transporters like H^+ ATPases [15]. Furthermore, high level of heavy metals lead to oxidative damage of plant tissues as a result of production of the reactive oxygen species [15]. Consequently, several toxicity symptoms in root and shoot systems may appear on plants in response to elevated heavy metal concentration [8]. Moreover, heavy metals accumulated within soil, may interfere in the food chain, leak into drinking water, and have a negative impact on human health, welfare live, and environment [2].

Basically, high concentrations of heavy metals are probably carcinogens for human and animals causing nucleic acid deformations and mutations [16]. For instance, Hwo [17] has showed that arsenic is carcinogenic for skin and deleterious for the cardiovascular system, whereas cadmium causes kidney damage as a result of its accumulation in kidney's tissues; similarly, mercury has harmful effects on the neurological system including uncontrolled muscles' movement, complete blindness, and malformation of new born children. In this regard, Padmavathiamma and Li [18] mentioned that lead exposure causes intensive damage to the nervous system which leads to poor understanding, unconcentrated memory, and losing of ability for learning and social collaboration, while arsenic also causes kidney diseases. Heavy metals are not easily biodegraded, threatening both environment and human health [9].

Increasingly, organic contaminants generating to environment from various sources such as usage of coal and fossil as a source of energy, military activities, as well as agriculture and industrial application; in fact, most organic pollutants are toxic and carcinogens [5]. Organic lipophilic matters have potential risk effect on human health, as they can interfere with food chain [19, 20]. Therefore, these contaminated soils are needed to be remediated because they usually cover large areas of the land.

Physical and Chemical Approaches of Remediation

Polluted sites can be cleaned up by physical, chemical, and biological techniques [21]. The physico-chemical strategies include soil excavation and storage, or transportation, washing, as well as chemical treatment [8]. These *ex situ* treatments of disturbed soil remove pollutants but at the same time they damage the soil microbial community. In addition, these approaches are very expensive [18, 22]. Glass [23] has summarized the high cost of physical and chemical methods of remediation, which have been estimated to approximately 75-425 US \$/ton for vitrification, and 20-200 US \$/ton for land filling and chemical treatments. In combination with the high economic cost, these applications generate hazardous substances behind them, containing heavy metals and additional pollutants, demanding further cure [24]. Furthermore, the physico-chemical remediation approaches are unfit for very large areas of contaminated sites such as mining sites, industrially and agriculturally polluted soil, or areas wasted by oil and gas operation. These procedures are improper for plant growth, beside they are also damageable for almost all soil biological activities and need huge amount of labour power [2, 25]. Additionally, *ex situ* remediation techniques modify and conversely harm physical, chemical, and biological traits of the treated soil [2]. Therefore, the environmental hazardous risks of *ex situ* remediation approaches combined together with the high costs have urged moving towards an alternative and cheaper technology to restore polluted sites [2]. Recently, researches have been oriented to an innovative field known as phytoremediation, in which plants and their associated soil microorganisms have been applied to remediate and improve disturbed soil's criteria [26, 27].

Phytoremediation: Application and Drawbacks

Phytoremediation is the use of plants and their rhizosphere-microbes to remove or immobilize contaminants from environment [27-29]. This technique is natural, green-clean, and an eco-friendly process to treat a wide variety of polluted soils including organic and inorganic wasted contaminants [5]. For instance, Pilon-Smits [5] has mentioned that inorganic pollutants which can be successfully remediated involve macronutrients such as nitrate and phosphate [30], essential trace elements like Cr, Ca, Mn, Mo, and Zn [31], non-essential ones such as Cd, Co, F, Hg, Se, Pb, V, and W [30, 32], and interestingly radioactive isotopes such as ^{238}U, ^{137}Cs, and ^{90}Sr [33-35].

As phytoremediation is a natural process, it depends on solar energy and does not require the transfer of contaminated soil for outside cleaning. Thereby, it is cheaper than other methods of remediations [23]. The same author scored that phytoextraction cost has been approximately 5-40 US \$/ton to get rid of phytomass products. In line with this, phytoremediation is less expensive (approximately 10-fold) than *ex situ* remediation [5]. In addition to the economical benefit of bioremediation, another benefit is minimizing exposure of human, wildlife, and environment to polluted products. Phytoremediation technology has received funding and is gaining popularity from many governments and environmental consultancy companies [5]. This green approach offer vegetation cover for a broad range of contaminated sites which produce extensive root system and high biomass, and therefore reduce and limit erosion [25, 36-39].

The application of phytoremediation, however, may be restricted by some drawbacks such as its slow process requiring many years to lower the pollutants concentrations into safe levels, and as it is a biological process, phytoremediation relies on various parameters including soil features, toxicity level, bioavailability of pollutants, as well as climatic conditions [5, 40]. Therefore, most of these factors should be well understood before phytoremediation would be widely acceptable as a commercial technique. Basically, to improve the efficacy of *in situ* remediation and to reduce the long time-periods required, it is tremendously important to ascertain the favourable plant traits to perform these purposes. The ideal plant material will be characterized by a fast growth, efficiency to accumulate and concentrate contaminants with the capacity to properly translocate them to their aboveground parts, high tolerance to a wide variety of pollutants, competitive species, extensive root production, high shoot biomass, high level of water and elements uptake, high transpiration rate, proper translocator or sequestrator of pollutants, as well as plants which can establish mutual relationship with rhizospheric soil microorganisms [2, 5, 41]. Examples of plant species which possess these traits are vetiver grass, hemp, sunflower, poplar, and willow [2, 5].

Plants and their related rhizosphere-microorganisms have been used in several types of phytoremediation, which can be subgrouped into the following categories as: phytoextraction, or disposal and accumulation of polluted compounds into aboveground plant tissues; phytostabilization, involving immobilization and sequestration of metals within soil and roots; rhizofiltration or absorption and adsorption of contaminants from water; phytodegradation and phytostimulation meaning degradation of organic pollutants into soil or within plant parts; and finally phytovolatilization, or releasing and volatilizing of polluted products via green plant organs from wasted soil into the atmosphere as less hazardous compounds [2, 5, 42-45].

Versatile Functions of Plants Used in Phytoremediation

Hyperaccumulators are plants which can intake and accumulate large quantities of heavy metals in their harvestable parts without the appearance of metal's toxicity symptoms [18]. Approximately 400 plant species have been reported as hyperaccumulators belonging to the following families: Asteraceae, Brassicaceae, Caryophyllaceae, Cyperaceae, Cunouniaceae, Euphorbiaceae, Fabaceae, Flacourtiaceae, Lamiaceae, Poaceae and Violaceae. The Environment Canada Agency has developed the Phytorem database, including approximately 750 plants, lichens, algae, fungi and bryophytes that have demonstrated the ability to tolerate, accumulate or hyperaccumulate a range of 19 different metals, including wild and cultivated plants [18]. Natural accumulator plants have shown concentrations of around 1% of Zn and Mn, 0.1% Ni, Co, Cr, Cu, Pb, and Al, 0.01% Cd and Se, and 0.001% Hg of their dry weight shoot biomass [46]. Labrecque et al. [47] showed that fast growing willows (*Salix viminalis* and *Salix discolour*) accumulated high contents of heavy metals such as mercury, copper, lead, nickel, and zinc in their roots and stems from soils treated with wastewater sludge. The same authors have suggested that the accumulation of heavy metals within roots and shoots have a significant role in the immobilization of heavy metals for several years, in contrast to the accumulation of heavy metals within the leaves because the leaves fall down annually and their heavy metals content return to soil. The heavy metals sequestration in the roots and stems provides a well route to decontamination of soils amended with heavy metals-containing wastewater sludge, and avoids pollutant contacting with the environment. Thus, this could allow to reduce the hazardous effect of pollution [47]. At high levels of heavy metals, both the essential and non-essential metals induce toxicity features and retardation of plant growth [15]. These toxicity symptoms originate from interactions between metals and sulphydryl groups of proteins which causes tackling and deformation in activity and structure of potential proteins, or structural substitution of necessary elements causing poisonous damages [48]. Moreover, elevated heavy metal concentrations accelerate the constitution of free radicals and reactivate oxygen species causing oxidative stresses [49]. However, some plant species can survive on contaminated soil; in fact, they possess various tolerance mechanisms allowing them to thrive on polluted soils [15].

Plant Heavy Metals Tolerance Mechanisms

Plants have many strategies to resist against heavy metal toxicities [15, 50], by reducing poisonous impacts of high concentrations of heavy metals via various mechanisms, such as controlling heavy metals intake, compartmentalization, translocation, and detoxification [50]. In this regard, heavy metal plant tolerance involve circumventing toxic metals building-up or producing proteins which can withstand against heavy metals poisoning [15]. Intercellular tolerance process include the role of plasma membrane in regulation of the uptake of heavy metals or speeding up the efflux pumping of metals out of the cytosol, producing heat shock proteins or metallothioneins which are involved in repairing of stress-disruption proteins, producing organic acids or amino acids which act as chelating agents, and compartmenting the metals within vacuoles [15].

Heavy metals have destructive influences on structure and function of plasma membrane involving oxidative damage of membrane proteins, limiting the activity of H^+ ATPases transporters, and modifying the structure and viscosity of membrane lipids [51]; thereby, hampering plasma membrane permeability [15]. However, plasma membrane has its own resistance to heavy metal toxicities such as metal hemeostasis; plasma membrane can control metals coming into and coming out to cytoplasm, regulating the active efflux pumping [15]. On the other hand, plasma membrane governs the metal tolerance through the potential functions of membrane transporters, such as CPx-ATPases, Nramps (natural resistance-associated macrophage protein), CDF (cation diffusion facilitator) [52], and ZIP (zinc transporter family) [53]. Nramp occurs in plants and microorganisms and performs an important role in transportation of heavy metals [52]. In addition, CDF, Nramp, and ATP-binding cassette play important functions in metal compartmentalization within vacuoles [6, 54-56].

Moreover, large groups of plants and other living organisms release heat shock proteins (HSPs) as a result of exposure to heavy metals [57, 58]. It was thought that HSPs offer functional and repairing role in heavy metals resistance mechanism [15]. Other types of proteins which have crucial role in metal toxicity resistance are the phytochelatins (PCs), which are glutamylcysteins oligopeptides containing glycine or other amino acids attached with a carboxyl terminus, in which γ-L-glutamyl-L-cysteine units (γ-Glu-Cys) are multiplied 2-11 times [41]. PCs play an important role in detoxification of Cd and As, but don't have an important effect on Zn, Ni, and selenite [15]. Metallothioneins (MTs) are sulphur rich proteins consisting of 60-80 amino acids which contain 9-16 cysteine, and having capacity to attach with heavy metals [59-61]. Interestingly, some encoding vital proteins genes contributing to metal tolerance have been recognized such as RAN1 [62], ATHMA3 [63], phytochelatin gene [64] and metallothionein genes [65, 66]. Intercellular resistance can also be contributed by dead cells of xylem and phloem containing cellulose, hemicellulose, and lignin which have capacity to trap metals reducing their toxicity impact into plant [67]. Moreover, heavy metals can be captured inside trichomes [68, 69].

On the other hand, plants achieve tolerance to the toxic effects of organic pollutants by uptake, transportation, volatilization, and sequestration inside vacuoles, or chemical modification involving oxidation, reduction, or hydrolysis and combination with glucose, glutathione, and amino acids [5, 27, 70, 71]. In addition, glutathione and glutathione-S-transferase play a functional action in degradation of organic pollutants such as most pesticides [71-73]. ATP-binding cassette (ABC) transporters have a crucial role in export of organic molecules out of plant cells and sequestration within vacuoles [70, 73]. Chelating agents including metallothionein, glutathione, phytochelatin [60], phytosiderophores [74], nicotinamine [75, 76], and organic acids like citrate, malate, and histidine [29, 76, 77] are produced by plant and are involved in trapping pollutants within vacuoles or exporting them to shoots.

Plant Root Exudates and Heavy Metal Tolerance

Plant roots release various organic acids consisting of citrate, lactate, and malate, as well as flavonoid compounds [78, 79], acting as carbon sources for soil microbial populations [42, 80, 81]. In response, some soil microorganisms produce plant growth hormones, increase the efficiency of water and nutrient uptake, and inhibit the action of other intruding soil microbes, consequently improving the nutritional and health conditions of plants [81]. Indeed, root-released lipophilic substances enhance pollutant solubility and movement, and stimulate the activities of biosurfactant-producing bacteria [82]. In this way, biosurfactants molecules promote the solubility of hydrophobic organic pollutants [83]. Additionally, plant roots together with soil microorganisms secrete degradable enzymes, using organic pollutants as substrates and speeding up the bioavailability of these contaminants [84]. Root exudates influence soil pH; in fact, soil acidification has strong impact on the availability of nutrients and toxic heavy metals [85]. The substructural epidermal root layers possess enormous Golgi apparatus and plasma membrane's vesicles which release siderophores products capturing a wide range of metals like iron, zinc, and arsenic, and therefore minimize their availability for root sorption [85]. Robinson and Anderson [86] suggested that root exudates ameliorate soil aeration through establishing avenues in soil for air and water exchange.

Fundamental Functions of Arbuscular Mycorrhizal Fungi in Phytoremediation

Arbuscular mycorrhizal fungi are ubiquitous soil microorganisms and are a vital component of the rhizosphere. Furthermore, AMF form a putative interaction with roots of approximately 80% of the terrestrial plants in nearly all ecosystems [87, 88]. Interestingly, AMF inhabit most of harsh conditions and climates [89], even though in deleterious soil involving heavy-metal contaminations [8, 90, 91]. AMF establish beneficial symbiotic relationships with plants and offer a physical bridge between soil and plant roots [87]. AMF constitute a large network of external hyphae within soil, these hyphae extending into the soil, reach nutrients in soil zones unavailable for direct plant uptake, and deliver these compounds to their host plants [92]. Therefore, AMF promote the nutrient supply to their hosts including phosphate, nitrogen, many micronutrients and other immobile molecule and water [92-99]. Moreover, AMF improve soil texture and reduce erosion through soil particles aggregation [100, 101], and increase the immobilization of heavy metals within soil by translocating metals into hyphae or roots; beside, AMF reduce metal moving from plants to soil and root-to-shoot translocation [102, 103].

Arbuscular mycorrhizal fungi can successfully colonize the root of some hyperaccumulator plant species and increase metal tolerance mechanisms and accumulations [25]. For example, AMF can establish symbiotic interaction with Ni-hyperaccumulator *Berheya coddi* [104], As-accumulator *Pteris vittata* [105, 106], *Cynodon dactylon* (hyperaccumulator for many heavy metals) [106], and *Thaspi praecox* [107]. Thus, this is indicating the role of mycorrhizal hyperaccumulator symbiosis in phytoextraction processes; unfortunately, most of hyperaccumulator plants produce small biomass and uptake a specific metal only [108]. Wang et al. [109] showed that AMF assist the ability of plants to uptake mineral nutrients, containing heavy metals. Moreover, AMF have the ability to reclaim the heavy metal contaminated soil to their host roots [110, 111] and were shown to stimulate plant resistance, reduce heavy metal toxicity impact, and promote plant growth under metal stress [25].

Soil structure and soil particles aggregation are important criteria of soil quality in stressful ecosystem [112]. Soil microorganisms help small soil granules to aggregate and accumulate into larger particles within root rhizosphere. This function have been carried out through particle adhesion with bacterial products [113], and branched AMF hyphal network [112]. In this regard, glomalin released from the AMF extraradical hyphae has a huge involvement in maintenance of soil aggregation [114]. Thereby, combination between AMF and rhizosphere bacteria contribute to soil aggregation and as a consequence, keep water availability [115, 116] and improve the restoration and revegetation of deleterious soil [116].

Interaction between Arbuscular Mycorrhizal Fungi and Rhizospheric Soil Microorganisms

Soil microbes and plant roots can sense each other via the release of certain molecules within soil which stimulate the activity of particular microbial populations to colonize plant root surface and inhibit other taxa. In fact, this specific plant-microbe communication is very complex and is governed by various biotic and abiotic factors. N-acyl-homoserine lactone is one of these signal molecules which is thought to be involved in quorum sensing and in the regulation symbiotic relations between root and soil bacteria [113]. AMF contribute to plant productivity and health by favouring nutrition, and providing tolerance against stressful conditions [117, 118]. AMF control the diversity and bioactivity of soil microorganisms within rhizosphere [9]. AMF colonization of root tissue can in some cases [119] lead to change in morphological structure of root tissues, but more widely demonstrated is a change in root exudatation resulting in a modification in rhizosphere's microbial communities structure and interactions with roots [113, 120-122], stimulation of plant growth, and increased plant resistance to harsh conditions [123, 124]. AMF have several impacts on soil bacteria and fungi [121, 123, 125-130]. It is also well established that AMF increase their host plant to escape diseases caused by deleterious microorganisms [131].

As AMF can acquire phosphorus from soil across their extraradical hyphal network, plant growth promoting rhizobacteria (PGPR) also increase the uptake of phosphorus [132]. Accordingly, Vivas et al. [133] demonstrated that AMF and PGPR isolated from polluted soil stimulate plant nutrition by improving N-fixation, producing plant growth hormones, and increasing P uptake. On line with this, *Pseudomonas* spp. as PGPR taxa benefit plant growth through phytohormone production, enzyme secretion, N-fixation, induced resistance to plant pathogens, production of antibiotic and other pathogen inhibitor, and immobilizing trace metal through the function of siderophores [81, 134-138].

Among antibiotics released by PGPR are acetyl-phluoroglucinol [139, 140] and phenazine [138]. Furthermore, PGPR are antagonistic against a variety of plant pathogens and showed increased nutrient uptake and phytostimulation [141-143]. PGPR have an impact on biological diversity of other microbial taxa of the rhizosphere [113]. Lippmann [144] recorded that *Pseudomonas* and *Acinetobacter* strains produce indole acetic acid (IAA) which increase the uptake of iron, zinc, magnesium, calcium, and potassium. Therefore, PGPR can be used as phytoremediative members which ameliorate plant traits including nutrition, health, and metal tolerance and detoxification [145].

Garbaye [146] have named the bacteria improving AMF hyphal growth and mycorrhizal establishment as mycorrhiza-helper bacteria (MHB). Vivas et al. [147, 148] demonstrated that inoculation of *Brevibacillus* spp. and *Glomus mosseae* isolated from Cd contaminated soil increase the nitrogen and phosphorus content and biomass of inoculated plants compared to control plants. Also, Barea et al. [113] concluded that microorganisms involving bacteria and AMF adapted to contaminated soils enhanced the mechanism of plant tolerance to heavy metals, and thus they have beneficial role in phytoremediation process. For these reasons, managing indigenous communities can contribute with plants to perform functions of vital importance in revegetation and remediation of disturbed soil [149, 150]. Hence, the interaction of AMF, PGPR, and MHB can be exploited as fascinating factors into biofertilizer and biocontrol application [151], as well as for phytoremediation implementation [9].

Heavy Metal Tolerance of Arbuscular Mycorrhizal Fungi

Arbuscular mycorrhizal fungi play a functional role in heavy metal resistance [152] and accumulation [153, 154], but they vary in their contribution and tolerance to heavy metal uptake and immobilization [37]. AMF own various mechanisms to protect themselves and their host against heavy metal toxicity damages [37, 155]. On one hand, immobilization and accumulation of heavy metals into fungal tissue and rhizosphere can be exploited to withstand metal toxicity especially in highly contaminated soil. Moreover, AMF can also act as a barrier in metal trafficking from root-to-shoot systems [156, 157]. Gaur and Adholeya [25] suggested that this is consequent to intra-cellular precipitation of metallic cations with phosphate groups. In this regard, Turnau showed higher accumulations of Cd, Ti, and Ba within fungal tissues than in host plant tissues. On the other hand, AMF can contribute several mechanisms contributing to adaptation to pollution stresses, including the crucial actions of cell wall's chitin [159], extraradical hyphae, and AMF-released proteins such as siderophore, metallothioneins, and phytochelatins [160]. Heavy metals such as Cu, Pb, and Cd can be trapped with cell wall structure including amino, hydroxyl, and carboxyl free radicals [161]. In addition, glomalin is an insoluble glycopeptide and chelating factor secreted by AMF [114] come out in the soil from AMF hyphae [162] and contribute to immobilization of metals and to the decreasing of bioavailability of metals [163], as well as help soil aggregation by adhering to soil particles [9]. The AMF-produced metal-binding glomalin have been extracted from contaminated soil under laboratory conditions [163]. Gonzalez-Chavez et al. [163] showed that 0.08 mg Cd, 1.12 mg Pb, and 4.3 mg Cu per gram glomalin was extracted from AMF grown in polluted soil. Also, one gram of glomalin containing 28 mg of Cu was obtained from *Gigaspora rosseae* [8], so it was hypothesized that glomalin can significantly contribute to capture heavy metals and sequester them within soil.

Over the role of fungal molecules such as chitin and glomalin in metal tolerance, the fungal hyphae offer a larger surface area than roots and a wide range of extension within soil, where they can grow and spread nearly elsewhere between soil granules where host root cannot grow [164, 165]. The extraradical AMF hyphae have higher affinity for metal attraction than host roots. For example, Chen et al. [166] observed that the concentration of P, Cu, and Zn were higher in the fungal tissue than in roots and shoots of maize when maize plants grown on modified glass bead compartment cultivation system with soil containing 0.80 of Cu mg/kg, 7.6 of Fe mg/kg, 3.6 of Mn mg/kg, and 0.63 of Zn mg/kg and colonized by *Glomus mosseae* and *G. versiforme*. For instance, *G. mosseae* and *G. versiforme* accumulated respectively approximately 1200 mg/kg and 600 mg/kg of Zn in their tissues while Zn concentrations in roots were lower than 100 mg/kg. Moreover, the variation in heavy metal accumulation within AMF tissues may concern the difference of AMF species morphology. For example, the concentration of nutrients and heavy metals in *G. mosseae* were higher than those in *G. versiforme* since *G. mosseae* produce higher external hyphae extension and lower spores count than those recorded in *G. versiforme* [166]. This is also confirming that heavy metals accumulate with high levels in AMF external hyphae more than in AMF spores [166]. As a consequence, AMF enhance root absorption efficiency, and thereby AMF assist in uptake/immobilize metals into the rhizosphere ecosystem and are significant contributors for heavy metal fixation within soil [25, 156]. In this regard, external AMF mycelia act as a biological sink of metals [103, 167] by its adsorption into cell wall or glomalin [114, 159]. Therefore, AMF diminish the exposure of plant to heavy metal poisoning [9]. A demonstration using scanning electron microscopy (SEM) and transition electron microscopy (TEM) revealed that AMF mycelia can attach Cu [168].

Furthermore, AMF vesicles were shown to play a vital role in metals detoxification [8]. Other indirect tolerance mechanisms have been shown, such as enhancing plant biomass which dilute heavy metal concentration within tissues, precipitation of polyphosphate particles, and compartmentalization within vesicles and vacuoles [103, 158].

AMF originating from deleterious soil enhance plant tolerance to heavy metal further than those isolated from non-contaminated soil [103, 169]. The distribution of AMF species in contaminated soil is also significantly lower than those found in non-polluted soil [170, 171]. In spite of the fact that frequency of AMF species and spores in metal polluted soil is low, this smaller amount of AMF spores in polluted soil does not appear to significantly restrict the establishment of mycorrhizal symbiosis [103, 169, 172, 173]. For instance, a low diversity of AMF species and spores was recorded into the rhizospheric soil of *Viola caluminaria* (yellow zinc violet plant) [174, 175] growing on highly Zn and Pb polluted soil. Spores of AMF isolated from Zn polluted soil have been shown to have a higher germination rate when compared with those isolated from non-contaminated sites when exposed to high Zn concentrations [176]. Investigation of the influence of heavy metal on spore germination and symbiotic formation was performed on only two isolates and indicated that *Glomus intraradices* (DAOM 1811602) is more resistant to elevated heavy metal concentration than *G. etunicatum* and that tolerance relies on differences in fungal genotype [177]. In addition, indigenous AMF populations can contribute to offset elevated heavy metal stresses for plant growth [178]. The same authors have reported that although the number of AMF spores decreased with long-term applications of sewage sludge containing high level of Zn, Cd, Cu, Ni, and Pb, the AMF species did not disappear totally from polluted and stressful conditions.

Several heavy metal-tolerant AMF species have been isolated from polluted soil. For examples, Gildon and Tinker [179] isolated *Glomus mosseae* from Zn contaminated soil which overcame approximately 100 mg/kg of Zn concentration, while Weissenhorn et al. [180] assessed mycorrhizal fungi from Cd polluted soil. Sambandan et al. [181] also recorded 15 AMF species from metal polluted sites in India, where the percentage of colonized roots ranged from 22 to 71% and approximately 622 AMF spores were counted per 100 gram soil, and *Glomus geosporum* was founded in all studied sites. Turnau and Haselwandter [118] found that approximately 70% of *Fragaria vesca* roots were infected by *Glomus mosseae* in Zn contaminated soil. Del Val et al. [178] identified *Glomus claroideum* from contaminated sludge soil.

High amount of heavy metals were shown to be accumulated in mycorrhizal structure [25]. Cavagnaro [182] showed that AMF can enhance Zn uptake by plant at low soil Zn concentration. Deram et al. [183] have revealed that AMF-mediated increase the accumulation of Cd in shoots of *Arrhenatherum elatius*; their results suggested that the significant role of AMF in Cd assimilation varied with season and soil Cd concentration. There was a significant positive correlation between shoot Cd concentration and arbuscules occurrence in roots of *Arrhenatherum elatius* (a perennial grass with high biomass and accumulate high concentration of Cd [184]) and a negative correlation between the frequency of AMF root colonization and soil Cd concentration. For example, arbuscules and vesicles of AMF disappeared when Cd concentration in soil reached their maximum value in May. This disappearance of AMF accompanied with the decreasing of Cd concentration in shoots confirm that AMF symbiosis have a dynamic impact in the uptake and accumulation of heavy metals by plants [183, 185]. The decreasing of Cd concentration in shoot system as a result of AMF disappearance indicates that seasonal variation of AMF lead to the protection of developing seeds from exposure to toxic injury of heavy metal in contaminated soil [183]. Furthermore, this seasonal of AMF colonization may be concerning with phenology of AMF species [183]. Moreover, an AMF inoculum composed of *Gigaspora margarita* ZJ37, *G. decipens* ZJ38, *Scutellospora gilmori* ZJ39, *Acaulospora* spp., and *Glomus* spp. have been shown to have a higher impact on phytoaccumulation of many heavy metals (Cu, Zn, Pb, and Cd) by maize plant when compared to an AMF inoculum involving only *Glomus caledonium* 90036 [186]. The consortia of AMF species contribute to higher uptake and transportation of heavy metals, as well as tolerance of heavy metals toxicity than single AMF species and therefore a mixture of AMF may be more effective in phytoremediation [187]. Mycorrhizal *Populus* trees (*Populus alba* and *Populus nigra*) inoculated with *Glomus mosseae* have been recorded as suitable Zn-accumulator plant; however, there was a variation in the ability of the two registered clones of willows to extract heavy metals from soil [188]. Although AMF increase the accumulation of Zn in leaves of poplar, AMF cause biochemical modification to improve the growth of plant, and therefore enhance plant tolerance to high Zn concentration [189]. The same authors revealed that changes in free putrescine (polyamine) has a significant role in the growth and development of higher plants [190] concentration in poplar inoculated with *Glomus mosseae* and grown on Zn polluted soil, where putrescine contribute in metal ion compartmentation [191].

On the other hand, Konzdroj et al. [192] indicated that mycorrhizal fungi originating from highly spoiled soils ameliorated Cd stabilization within sites which were planted with pines. For example, Sudová and Vosátka [193] recorded that maize plants grown on gamma sterilized field substrate from lead polluted waste disposal site and inoculated with *G. intraradices* (isolate BEG 75 from non-polluted soil) had the lower P concentration in their shoots than maize inoculated with *G. intraradices* (isolates PH5-OS and PH5-IS from lead contaminated sites). The non-inoculated maize plants had more than double Pb concentration in their shoot biomass as compared with mycorrhizal plants, without any significant variations between the effects of *G. intraradices* isolates. Increasing Pb accumulation in mycorrhizal colonized roots together with lowing Pb levels in shoots of mycorrhizal plants may confirm the significant role of intraradical fungal components in heavy metal sequestration within the roots, either on cell walls or intracellullarly [193]. In addition, extraradical mycelia of AMF contribute in heavy metal immobilization [156]. Similarly, Janousková et al. [194] recorded higher accumulation of Cd in extraradical hyphae of AMF than in plant roots and lower Cd poisoning in mycorrhizal plant than non-mycorrhizal plants. Hence, internal and external hyphal structures of AMF have a significant function in alleviating of heavy metal toxicity and increasing the heavy metal immobilization within soil [193].

Since AMF decrease heavy metal accumulation in plant shoots, AMF offer a protection role for their host against heavy metal toxicity and resulting in high shoot yields [195-197]. For example, Wang et al. [198] recorded that mycorrhizal symbiosis in *Zea mays* and *Acaulospora mellea* accumulated low amount of Cu which had no or less toxicity and did not cause damageable injury of plants, and thus this lead to higher plant shoot yields. On the other hand, AMF increase soil pH, change the concentration of soil organic acids such as malic acid, citric acid, and oxalic acid, and influence on exudation of carbohydrate compounds [156]. These modifications may have a significant effect on metal availability [198]. Further, Vivas et al. [199] have also observed that the co-inoculation of *Trifolium repens* with the AMF *G. mosseae* and the bacteria *Brevibacillus brevis* promoted plant growth, mineral nutrition uptake, and reduced nickel uptake. This suggests that these synergistic effects of AMF bacteria can be exploited in biotechnological approaches to increase the efficiency of phytostabilization.

In a phytagel experiment, *G. intraradices* colonized carrot roots at high content of Cd and Zn in M media, carrot roots were hyperaccumulator of Cd (90 µg/g Cd) and accumulator of Zn (550 µg/g Zn) [200]. AMF hyphae could transfer Zn to their host roots by the same transport pathway of phosphorus [201, 202]. Arbuscules of AMF increase the surface connection area between AMF and roots, and thereby provide a significant role in Zn uptake to roots [92]. The well establishment of arbuscules and development of AMF within root increase metal translocation in plant shoot [104]. Although Cd is non-essential nutrient, it can be translocated and accumulated within plants through the manganese and zinc transport systems [200]. Beside the role of metal-lothioneins and phytochelatins in capture of heavy metals in their cysteinyl radicles, heavy metal can be accumulated in carrot roots infected with *G. intraradices* as a result of heavy metal saturation in vegetation after long period of metal exposure [200]. The different strategies of detoxification are very important for successful thriving of mycorrhizal plant on heavy metal polluted soil and ameliorating the restoration of contaminated sites [186].

Phytoextraction

Phytoextraction is a biotechnological approach in which contaminants are uptake from soil by plants and stored within their harvestable tissues. Therefore, ideal plants for this biological approach should be more tolerant and adapted to heavy metal stress, good metal accumulators, producing high biomass, an extensive root system, and fast growing [5, 203]. When the harvestable plant tissues contain relative quantities of heavy metals, they can be re-extracted in a new fascinating technique called phytomining [32, 204]. Nevertheless, in order to eliminate the environmental risks of plant residues containing heavy metals, the harvested plant parts can be used as a source of energy by combustion and stored in very small amount of dry matter [41, 203]. However, specific burning techniques are required to prevent metal losing with smokes . Beside, the use of a phytoremediative plant cover in forestry industry and biofuel production are promising environmental routes to reduce the limitations of phytoextraction [206, 207].

However, the most important factor restricting the application of phytoaccumulation is metal bioavailability. The chemical and physical features of soil together with other environmental conditions reduce the feasibility of contaminants' movement within soil [5]. The metal bioavailability is related to the solubility of this one in the soil solution. It also depends on the metal status in the soil. Metal would not be bio-available when they form complexes with soil particles or when they are precipitated as carbonate or hydroxides or phosphates [56]. Yet, induced accumulation of metals can be exploited by using synthetic chelating agents such as ethylene diamine tetraacetic acid (EDTA), nitrilotriacetic acid (NTA), thiosulphate, and thiocynate to increase the capacity of metal extraction from contaminated soil [208-211]. Some environmental hazard associated with the use of these synthetic chelators and involving metal leaching to underground water can happen [212]. Moreover, some chelators like EDTA are not easily degraded and are still present within soil after remediation, adding an additional pollutant into the environment [207]. Furthermore, many of the chelators are used as sodium salts, and it is known that plant growth has been reduced as a result of high Na concentration into soil [207]. Also, most synthetic chelating agents increase the solubility and availability of many metals other than those targeted by the remediation; as a consequence, new phytotoxicities coming from these non-target metals have been shown [86]. Therefore, more research would be needed to restrict the drawbacks of synthetic chelating agents' application.

Obviously, AMF can be applied to increase the efficiency of phytoaccumulation via their direct and indirect effects in heavy metals uptake and accumulation, and in plant biomass production [25, 200]. In addition, AMF enhance the plant contact area with soil through their extended hyphae and increase the root uptake area up to 47-fold [92]. Mycorrhizal infection stimulate metal translocation to root of lettuce [102], while other results have revealed that AMF speed up the accumulation of trace metals within shoot systems of legume plants [154, 213, 214]. To this regard, similar observations were recorded with other plants [200, 215-219]. The combined inoculation of an AMF and a *Penicillium* isolate plus the application of chistosan (a chelating agent) promoted the shoot and root growth of *Elsholtizia splendens* and speed up the translocation of Zn, Pb, and Cd, but not of Cu to the shoots [108]. Chitosan is chelating agent for ions of heavy metals because of its free amine function, which combine with cations in polluted soils [220]. Weng et al. [221] recorded that chitosan increase the accumulation of Cu and Pb in roots and shoots of *Elsholtizia splendens*, and together with AMF increase the concentration of Zn and Pb in the roots. Chitosan has low molecular weight, is water soluble, biodegradable and does not hamper plant growth and AMF establishment [108, 222, 223].

Furthermore, chitosan can be degraded after phytoremediation application and can no longer chelate heavy metals; thereby, it can be used as eco-friendly chelator to increase the role of AMF in phytoextraction of heavy metal polluted soil [108].

Phytostabilization

Stabilization of heavy metals within soil have been carried out by precipitation of metals into rhizosphere, adsorption onto root surface and soil particles, or absorption and accumulation within roots [5, 8]. Heavy metal immobilization has been shown with plants and their associated microorganisms [45]. Phytosequestration of trace metals into soil result in the restriction of metal spreading, leaching to underground water, and finally reduce erosion [5]. Also, different authors [224, 225] have concluded that phytosequestration is suitable and an alternative route whenever phytoextraction is not feasible. As some trees produce high biomass, deep and branched root system, high transpiration rate, and provide metal-organic matters to soil, they are consequently suitable for immobilization purposes [225-227]. Mertens et al. [228] have showed that trees which reduce soil pH and minimize the metal transition from root to shoot have an interesting function for phytostabilizative goal. By essence, metallophyte plants own a wide variability to thrive on highly contaminated soil, as they have different metal tolerance procedures [229]. Yoon et al. [230] reported 17 plant species which can survive on contaminated sites and show high metal concentration within roots and low metal translocation value from roots to shoots, and have significant applications for phytostabilization.

Immobilizations of contaminants within soil reduce the exposure of human and the environment to hazardous effects of pollutants [18]. Phytostabilization is most proper for soil with high organic and heavy metal contents, and is suitable for cure of a huge range of contaminated sites [231, 232]. Phytostabilization is less costly and easier for application than other procedures of phytoremediation [232, 233]. Phytostabilization has another advantage that make this procedure applicable as an eco-friendly method of restoration of polluted sites, such as there is no need to rid out of shoot systems as a hazardous waste [234]. The mechanisms of immobilization involve chelation of contaminants via root exudation, absorption, adsorption, and accumulation within roots, compartmentation inside vacuoles or combining with cell wall components, precipitation within rhizosphere, and reducing xylem transportation which reduce the translocation of contaminants from root to shoot systems [18].

To increase the efficacy of phytostabilization of heavy metal polluted soil, some amendments are used to increase the insolubility of metals and cause them unavailable for plant uptake; hence, heavy metals cannot interfere with food chain [231, 232, 235]. Phosphates as multi anions and organic compounds such as compost are one of these amendments that increase the immobilization and precipitation of heavy metals. Since phosphate raise metal adsorption by anion-induced negative charge and metal precipitation [18, 237], and organic compost enhance soil pH, these amendments may improve physical and chemical soil properties, increase heavy metal immobilization, and play an important role in the restoration of metal polluted soil [236]. Although phosphate addition increases the availability of arsenic from mine tailings, phosphate is required to facilitate the revegetation of mine tailing sites [18].

Furthermore, AMF can infect metalophyte plants and increase their ability to survive on highly contaminated soil by avoiding trace metal absorption [37, 103, 202, 229, 238-240]. Obviously, trace metals immobilization have been succeeded on acidic mine tailings [241], wasteful wood sites [242], and sheep waste areas [207]. Additionally, increase of the accumulation of metals within rhizosphere lead to add further organic matter to soil and establish vegetation cover on highly polluted soil [207]. The reduction of mycorrhizal colonization rate as a result of high metal availability may offer a mechanism for the restriction of heavy metal uptake and increase the fixation of heavy metals within soil [243]. In soil with high Cd and Zn contents, the hyphae of *G. mosseae* provide a barrier for heavy metal translocation and reduce the heavy metal uptake by *Phseolus vulgaris*, and this is due to the capture of heavy metal in hyphae-released slime [244, 245]. Furthermore, AMF have ability to change the forms of the contaminants; for example, AMF can reduce arsenate to arsenite and remove arsenite from their hyphae [246, 247]. Thereby, AMF can significantly uptake macronutrients and exclude heavy metals [248]. Obviously, AMF symbiosis can create more adaptation to heavy metal polluted condition by various mechanisms that increase the immobilization of heavy metal and a consequent to improve the phytostabilization technique [8].

Conclusion

Phytoremediation is an attracting technology to dispose or immobilize contaminants in derelict soils. One reason is because phytoremediation is an *in situ* approach avoiding the transportation of contaminated soil for *ex situ* treatment detoxification, making it an innovative, inexpensive, and popular approach. To enhance the efficacy of phytoremediative techniques and to minimize the long period of time required for cleanup, tolerant plants and their effective associated rhizospheric microorganisms are fundamental tools to reach this important environmental goal.

The use of AMF isolated from contaminated sites is a promising tool either in phytoaccumulation or phytosequestration techniques. The interaction between AMF and plant roots can be established in almost habitats and as a reason that AMF exist in heavy metal polluted soil, AMF can provide advantages to facilitate the growth of plants in polluted soil. Extraradical hyphae of AMF increase the ability of the roots to access unavailable nutrient and enhance water uptake. Hence, AMF play an important role in enhancing the biomass of their host plants. Since AMF release metal chelating agents such as glomalin, metallothionien, organic acid, and phytochelatin, AMF can increase the immobilization and sequestration of heavy metal within soil. AMF can further increase the reduction of metal ions by the potential role of the specific plasma membrane metal reductases [249]. Therefore, AMF filter the entrance of heavy metals to plant shoots and increase the avoiding of heavy metal toxicity. Moreover, indigenous AMF isolated from polluted site show more resistance and adaptation to deleterious conditions than those isolated from non-contaminated soils. Indigenous AMF isolated from polluted sites and the role of their vacuoles, vesicles, and arbuscules in storage and translocation of heavy metals, add more advantages of AMF in phytoremediation and restoration of polluted sites. The protective role of AMF at heavy metal toxicity tolerance relies on diverse biotic and abiotic factors involving diversity of plant, fungal, and microbial species and varieties, nutritional conditions of mycorrhizal plant, healthy states of plant, nutrients founding within soil, qualitative features of soil, and metal bioavailability and concentration [152, 185, 187, 202, 245, 250-253].

References

1. Gremion, F., Chatzinotas, A., Kaufmann, K., Sigler, W. et al. *FEMS Microbiology Ecology* 2004, 48:273-283.
2. Khan, A.G. *Journal of Trace Elements in Medicine and Biology* 2005, 18:355-364.
3. Kapoor, R., Chaudhary, V. and Bhatnagar, A. *Mycorrhiza* 2007 17:581-587.
4. Nriagu, J.O. *Nature* 1979, 279:409-411.
5. Pilon-Smits, E. *Annual Review of Plant Biology* 2005, 56:15-39.
6. Singh, O.V., Labana, S., Pandey, G., Budhiraja, R. et al. *Applied Microbiology and Biotechnology* 2003, 61:405-412.
7. Holleman, A. and Wiberg, E. *Lehrbuch der Anorganischen Chemie.* Berlin, 1985.
8. Göhre, V. and Paszkowski, U. *Planta* 2006, 223:1115-1122.
9. Khan, A.G. *Journal of Zhejiang University Science B* 2006, 7:503-514.
10. Mertz, W. *Science* 1981, 213:1332-1338.
11. Hall, J.L. and Williams, L.E. *J Exp Bot* 2003, 54:2601-2613.
12. Krupa, P. and Piootrowska-Seget, Z. *Polish Journal of Environmental Studies* 2003, 12:723-726.
13. Babich, H. and Stotzky, G. *Environmental Research* 1985, 36:111-137.
14. Giller, K., Witter, E. and McGrath, S. *Soil Biology and Biochemistry* 1998, 30:1389-1414.
15. Hall, J.L. *J Exp Bot* 2002, 53:1-11.
16. Knasmuller, S., Gottmann, E., Steinkellner, H., Fomin, A. et al. *Mutation Research* 1998, 420:37-48.
17. WHO. Health and environment in sustainable development. Geneva, 1997.
18. Padmavathiamma, P. and Li, L. *Water, Air, and Soil Pollution* 2007, 184:105-126.
19. Reilley, K., Banks, M.K. and Schwab, A.P. *J Environ Qual* 1996, 25:212-219.

20. Henner, P., Schiavon, M., Morel, J.L. and Lichtfouse, E. *Analysis* 1997, 25:M56-M59.
21. Mc Eldowney, S., Hardman, D.J. and Waite, S. *In:* Pollution Ecology and Biotreatment Technologies. Mc Eldowney, S., Hardman, J. and Waite, S. (eds), Longman, Singapore Publishers, Singapore, 1993.
22. Chaudhry, Q., Blom-Zandstra, M., Gupta, S. and Joner, E.J. *Environ Sci Pollut Res* 2005, 12:34-48.
23. Glass, D.J. U.S. and International Markets for Phytoremediation 1999-2000, Needham MA, D. Glass Assoc, 1999.
24. Williams, G.M. *In:* Land Disposal of Hazardous Waste, Engineering and Environmental Issues. Gronow, J.R., Schofield, A.N. and Jain, R.K. (eds), Horwood Ltd, Chichester, UK, 1988.
25. Gaur, A. and Adholeya, A. *Current Science* 2004, 86:528-534.
26. Brooks, R.R. *In:* Plants that hyperaccumulate heavy metals. Brook, R.R. (ed.), CAB International, Wallingford, UK, 1998.
27. Salt, D.E., Smith, R.D. and Raskin, I. *Annu Rev Plant Physiol Plant Molecular Biology* 1998, 49:643-668.
28. Raskin, I., Kumar, P.B.A.N., Dushenkov, S. and Salt, D.E. *Current Opinion in Biotechnology* 1994, 5:285-290.
29. Salt, D.E., Blaylock, M., Kumar, N.P.B.A., Dushenkov, V. et al. *Nat Biotech* 1995, 13:468-474.
30. Horne, A.J. *In:* Phytoremediation of Contaminated Soil and Water. Terry, N., Bañuelos, G. and Raton, B. (eds), Lewis, 2000, 13-40.
31. Lytle, C.M., Lytle, F.W., Yang, N., Qian, J.H. et al. *Environmental Science and Technology* 1998, 32:3087-3093.
32. Blaylock, M.J. and Huang, J.W. *In:* Phytoremediation of Toxic Metals: Using Plants to Clean up the Environment. Raskin, I. and Ensley, B.D. (eds), Wiley, New York, 2000, 53-70.
33. Dushenkov, S. and Kapulnik, Y. *In:* Phytoremediation of Toxic Metals: Using Plants to Clean up the Environment. Raskin, I. and Ensley, B.D. (eds), Wiley, New York, 2000, 89-106.
34. Negri, M.C. and Hinchman, R.R. *In:* Phytoremediation of Toxic Metals: Using Plants to Clean up the Environment. Raskin, I. and Ensley, B.D. (eds), Wiley, New York, 2000, 107-132.
35. Dushenkov, S. *Plant and Soil* 2003, 249:167-175.
36. Zak, J.C. and Parkinson, D. *Can J Bot* 1982, 60:2241-2248.
37. Leyval, C., Turnau, K. and Haselwandter, K. *Mycorrhiza* 1997, 7:139-153.
38. Glick, B.R., Patten, C.L., Holguin, G. and Penrose, D.M. Biochemical and genetic mechanisms used by plant growth-promoting bacteria. Imperial College Press, London, UK, 1999.
39. Burd, G.I., Dixon, D.G. and Glick, R.R. *Can J Microbiol* 2000, 46:237-245.
40. Flechas, F.W. and Latady, M. *Adv Biochem Engin/Biotechnol* 2003, 78:172-185.
41. Peuke, A.D. and Rennenberg, H. *EMBO Reports* 2005, 6:497-501.
42. Terry, N., Zayed, A., Pilon-Smits, E. and Hansen, D. *In:* Will Plants Have a Role in Bioremediation? Proc. 14th Annu. Symp., Curr. Top. Plant Biochem, Physiol. Mol. Biol, Univ. Missouri, Columbia, April 19-22, 1995, 63-64.
43. Newman, L.A., Strand, S.E., Choe, N., Duffy, J. et al. *Environ Sci Technol* 1997, 31:1062-1067.

44. Chaudhry, T.M., Hayes, W.J., Khan, A.G. and Khoo, C.S. *Australian Journal Ecotoxicol* 1998, 4:37-51.
45. Berti, W.R. and Sd, C. *In:* Phytoremediation of Toxic Metals: Using Plants to Clean up the Environment. Raskin, I. and Ensley, B.D. (eds), Wiley, New York, 2000, 71-88.
46. Baker, A.J.M. and Brooks, R.R. *Biorecovery* 1989, 1:81-126.
47. Labrecque, M., Teodorescu, T. and Daigle, S. *Plant and Soil* 1995, 171:303-316.
48. Van Assche, F. and Clijsters, H. *Cell and Environment* 1990, 13:195-206.
49. Dietz, K-J. and Krämer, U. *In:* Heavy metal stress in plant: from molecular to ecosystems. Prasad, M.N.V. and Hagemeyer, J. (eds), Springer-Verlag, Berlin, 1999, 73-97.
50. Clemens, S. *Planta* 2001, 212:475-486.
51. Meharg, A.A. *Physiologia Plantarum* 1993, 88:191-198.
52. Williams, L.E., Pittman, J.K. and Hall, J.L. *Biochim Biophys Acta* 2000, 1465:104-126.
53. Guerinot, M.L. *Biochimica et biophysica Acta* 2000, 1465:190-198.
54. Paulsen, I.T. and Saier, M.H. Jr. *J Membr Biol* 1997, 156:99-103.
55. Rea, P.A. *J Exp Bot* 1999, 50:895-913.
56. Clemens, S., Palmgren, M.G. and Krämer, U. *Trends Plant Sci* 2002, 7:309-315.
57. Vierling, E. *Annu Rev Plant Physiol Plant Molecular Biology* 1991, 42.
58. Lewis, S., Handy, R.D., Cordi, B., Billinghurst, Z. et al. *Ecotoxicology* 1999, 8:351-368.
59. Rauser, W.E. *Cell Biochem Biophys* 1999, 31:19-48.
60. Cobbett, C.S. and Goldsbrough, P.B. *In:* Phytoremediation of Toxic Metals: Using Plants to Clean up the Environment. Raskin, I. and Ensley, B.D. (eds), Wiley, New York, 2000, 247-271.
61. Cobbett, C. and Goldsbrough, P. *Annu Rev Plant Physiol Plant Biology* 2002, 53:195-182.
62. Hirayama, T., Kieber, J.J., Hirayama, N., Kogan, M., Guzman, P., Nourizadeh, S. et al. *Cell* 1999, 97:383-393.
63. Gravot, A., Lieutaud, A., Verret, F., Auroy, P., Vavasseur, A. and Richaud, P. *FEBS Letters* 2004, 561:22-28.
64. Clemens, S., Kim, E.J., Neumann, D. and Schroeder, J.I.. *EMBO Journal* 1999, 18:3325-3333.
65. Prasad, M.N.V. *In:* Phytoremediation of Toxic Metals: Using Plants to Clean up the Environment. Raskin, I. and Ensley, B.D. (eds), Wiley, New York, 2000, 51-72.
66. Goldsbrough, P. *In:* Phytoremediation of contaminated soil and water. Terry, N. and Banuelos, G. (eds), CRC Press LLC, 2000, 221-233.
67. Monni, S., Bucking, H. and Kottke, I. *Micron* 2002, 33:339-351.
68. Choi, Y.E., Harada, E., Wada, M., Tsuboi, H. et al. *Planta* 2001, 213:45-50.
69. Kupper, H., Jie Zhao, F. and McGrath, S.P. *Plant Physiol* 1999, 119:305-312.
70. Meagher, R.B. *Current Opinion Plant Biol* 2000, 3:153-162.
71. Dietz, A.C. and Schnoor, J.L. *Environ Health Perspect* 2001, 109:163-168.
72. Edwards, R., Dixon, D.P. and Walbot, V. *Trends Plant Sci* 2000, 5:193-198.
73. Dixon, D.P., Lapthorn, A. and Edwards, R. *Genome Biol* 2002, 3:REVIEWS3004.
74. Higuchi, K., Suzuki, K., Nakanishi, H., Yamaguchi, H. et al. *Plant Physiol* 1999, 119:471-480.

75. Stephan, U.W., Schmidke, I., Stephan, V.W. and Scholz, G. *BioMetals* 1996, 9:84-90.
76. von Wiren, N., Klair, S., Bansal, S., Briat, J.-F. et al. *Plant Physiol* 1999, 119:1107-1114.
77. Kupper, H., Mijovilovich, A., Meyer-Klaucke, W. and Kroneck, P.M.H. *Plant Physiol* 2004, 134:748-757.
78. Ensley, B.D. *In:* Phytoremediation of Toxic Metals: Using Plants to Clean up the Environment. Raskin, I. and Ensley, B.D. (eds), Wiley, New York, 2000, 3-12.
79. Hutchinson, S.L., Schwab, A.P. and Banks, M.K. *In:* Phytoremediation: Transformation and Control of Contaminants. McCutcheon, S.C. and Schnoor, J.L. (eds), Wiley, New York, 2003, 355-86.
80. Bowen, G.C. and Rovira, A.D. *In:* Plant Roots—The Hidden Half. Waisel, Y., Eshel, A. and Kaffkafi, U. (eds), Marcel Dekker, New York, 1991, 641-69.
81. Kapulnik, Y. *In:* Plant Roots—The Hidden Half. Waisel, Y., Eshel, A. and Kaffkafi, U. (eds), Marcel Dekker, New York, 1991, 769-81.
82. Siciliano, S.D. and Germida, J.J. *Environ Rev* 1998, 6:65-79.
83. Volkering, F., Breure, A.M. and Rulkens, W.H. *Biodegradation* 1997, 8:401-417.
84. Wolfe, N.L. and Hoehamer, C.F. *In:* Phytoremediation: Transformation and Control of Contaminants. McCutcheon, S.C. and Schnoor, J.L. (eds), Wiley, New York, 2003, 159-87.
85. Meagher, R.B. and Heaton, A.C.P. *J Ind Microbiol Biotechnol* 2005, 32:502-513.
86. Robinson, B.H. and Anderson, C.W.N. *In:* Phytoremediation methods and reviews. Willey, N. (ed.), Humana Press, Totowa, NG, 2006, 453-466.
87. Barea, J.M. and Jeffries, P. *In:* Mycorrhiza structure, function, molecular biology and biotechnology. Hock, B. and Varma, A. (eds), Springer-Verlag, Heidelberg, Germany, 1995, 521-559.
88. Brundrett, M.C. *New Phytologist* 2002, 154:275-304.
89. Chaudhry, T.M. and Khan, A.G. *In:* Biotechnology of microbes and sustainable utilization. Rajak, R.C. (ed.), Scientific Publishers, Jodhpur, India, 2002, 270-279.
90. Estaún, V., Savé, R. and Biel, C. *Appl Soil Ecol* 1997, 6:223-229.
91. Enkhtuya, B., Rydlová, J. and Vosátka M. *Appl Soil Ecol* 2002, 14:201-211.
92. Smith, S.E. and Read, D.J. Mycorrhizal symbiosis. 2nd ed., Academic Press, San Diego, CA, 1997.
93. Li, X.L., George, E. and Marschner, H. *Soil Plant Soil* 1991, 136:41-48.
94. Li, X.-L., George, E. and Marschner, H. *New Phytologist* 1991, 119:397-404.
95. Jakobsen, I., Abbott, L.K. and Robson, A.D. *New Phytologist* 1992a, 120:371-380.
96. Jakobsen, I., Abbott, L.K. and Robson, A.D. *New Phytologist* 1992b, 120:509-516.
97. Li, X.L., George, E., Marschner, H. and Zhang, J.L. *Canadian Journal of Botany* 1997 75:723-729.
98. Vivas, A., Marulanda, A., Gómez, M., Barea, J.M. et al. *Soil Biol Biochem* 2003, 35:987-996.
99. Yao, Q., Li, X., Weidang, A. and Christie, P. *Eur J Soil Biol* 2003, 39:47-54.
100. Rillig, M.C. and Steinberg, P.D. *Soil Biol Biochem* 2002, 34:1371-1374.
101. Steinberg, P.D. and Rillig, M.C. *Soil Biol Biochem* 2003, 35:191-194.
102. Dehn, B. and Schuepp, H. *Agriculture, Ecosystems and Environment* 1989, 29:79-83.
103. Kaldorf, M., Kuhn, A.J., Schröder, W.H., Hildebrandt, U. et al. *Journal of Plant Physiology* 1999, 154:718-728.

104. Turnau, K. and Mesjasz-Przybylowicz, J. *Mycorrhiza* 2003, 13:185-190.
105. Al Agely, A., Sylvia, D.M. and Ma, L.Q. *J Environ Qual* 2005, 34:2181-2186.
106. Leung, H.M., Ye, Z.H. and Wong, M.H. *Environmental Pollution* 2006, 139:1-8.
107. Vogel-Mikus, K., Drobne, D. and Regvar, M. *Environmental Pollution* 2005 133:233-242.
108. Wang, F.Y., Lin, X.G. and Yin, R. *Environmental Pollution* 2007, 147:248-255.
109. Wang, F., Lin, X. and Yin, R. *Plant and Soil* 2005, 269:225-232.
110. Meharg, A.A. and Cairney, J.W.G. *Advance in Ecological Research* 1999, 30:70-112.
111. Marques, A.P.G.C., Oliveira, R.S., Rangel, A.O.S.S. and Castro, P.M.L. *Chemosphere* 2006, 65:1256-1263.
112. Miller, R.M. and Jastrow, J.D. *In:* Arbuscular mycorrhizas: Physiology and Functions. Kapulnik, Y. and Douds Jr (eds), Kluwer Academic Publishers, Dordrecht, The Netherlands, 2000, 3-18.
113. Barea, J.M., Pozo, M.J., Azcon, R. and Azcon-Aguilar, C. *Journal of Experimental Botany* 2005, 56:1761-1778.
114. Wright, S.F. and Upadhyaya, A. *Plant and Soil* 1998, 198:97-107.
115. Bethlenfalvay, G.J., Cantrell, I.C., Mihara, K.L. and Schreiner, R.P. *Biology and Fertility of Soils* 1999, 28:356-363.
116. Requena, N., Perez-Solis, E., Azcon-Aguilar, C., Jeffries, P. et al. *Applied and Environmental Microbiology* 2001, 67:495-498.
117. Clark, R.B. and Zeto, S.K. *J Plant Nutri* 2000, 23:867-902.
118. Turnau, K. and Haselwandter, K. *In:* Mycorrhizal Technology: from Genes to Bioproducts. Gianinazzi, S. and Schuepp, H. (eds), Birkhauser, Basel, 2002, 137-149.
119. Chapdelaine, A., Dalpé, Y., Hamel, C. and St-Arnaud, M. *In:* Mycorrhiza works: Applications and real case field studies. Feldmann, F., Kapulnik, Y. and Baar, J. (eds), COST 870 Meeting and the German Phytomedical Society, 2009 (in press).
120. Pfleger, F.L. and Linderman, R.G. *Mycorrhizae and Plant Health,* Symposium Series, The American Phytopathological Society, St Paul, Minnesota, USA, 1994, p344.
121. Marschner, P. and Baumann, K. *Plant and Soil* 2003, 251:279-289.
122. Lioussanne, L., Jolicoeur, M. and St-Arnaud, M. *Soil Biology and Biochemistry* 2008 (in press).
123. Paulitz, T.C. and Linderman, R.G. *New Phytologist* 1989, 113:37-45.
124. Lynch, J.M. The Rhizosphere. John Wiley and Sons, West Sussex, UK, 1990.
125. Hamel, C., Vujanovic, V., Jeannotte, R., Nakano-Hylander, A. et al. *Plant and Soil* 2005, 268:75-87.
126. Yergeau, E., Vujanovic, V. and St-Arnaud, M. *Microbial Ecology* 2006, 52:104-113.
127. Filion, M., St-Arnaud, M. and Fortin, J.A. *New Phytologist* 1999, 141:525-533.
128. Filion, M., St-Arnaud, M. and Jabaji-Hare, S.H. *Phytopathology* 2003, 93:229-235.
129. St-Arnaud, M. and Elsen, A. *In:* In vitro culture of mycorrhizas Soil biology Series. Declerck, S., Strullu, D.-G. and Fortin, J.A. (eds), Springer-Verlag, Berlin, 2005, 217-231.
130. Scheffknecht, S., St-Arnaud, M., Khaosaad, T., Steinkellner, S. et al. *Canadian Journal of Botany* 2007, 85:347-351.

131. St-Arnaud, M. and Vujanovic, V. *In:* Mycorrhizae in crop production. Hamel, C. and Plenchette, C. (eds), Haworth Press, Binghampton, NY, 2007, 67-122.

132. Rodríguez, H. and Fraga, R. *Biotech Advances* 1999, 17:319-339.

133. Vivas, A., Marulanda, A., Ruiz-Lozano, J., Barea, J. et al. *Mycorrhiza* 2003, 13:249-256.

134. Weller, D.M. *In:* Plant Roots: The Hidden Half. Waisel, Y., Eschel, A. and Kafkafi, U. (eds), Dekker, New York, 1988, 769-781.

135. Glick, B.R. *Can J Microbiol* 1995, 41:109-117.

136. Kamnev, A.A. and Van der, L. *Biosci Rep* 2000, **20**:239-258.

137. Zhang, S., Ms, R. and Kloepper, J.W. *Biol Control* 2002, 23:79-86.

138. Chin-A-Woeng, T.F.C., Bloemberg, G.V. and Lugtenberg, B.J.J. *New Phytologist* 2003, 157:503-523.

139. Landa, B.B., Mavrodi, D.M., Thomashow, L.S. and Weller, D.M. *Phytopathology* 2003, 93:982-994.

140. Picard, C., Frascaroli, E. and Bosco, M. *FEMS Microbiology* 2004, 49:207-215.

141. Persello-Cartieaux, F., Nussaume, L. and Robaglia, C. *Plant, Cell and Environment* 2003, 26:189-199.

142. Barea, J.M., Azcon, R. and Azcon-Aguilar, C. *In:* Plant surface microbiology. Varma, A., Abbott, L., Werner, D. and Hampp, R. (eds), Springer-Verlag, Heidelberg, Germany, 2004, 351-371.

143. Zahir, Z.A., Arshad, M. and Frankenberger, W.T. *Advances in Agronomy* 2004, 81:97–168.

144. Lippmann, B., Leinhos, V. and Bergmann, H. *Angew Bot* 1995, 69:31-36.

145. Mayak, S., Tirosh, S. and Glick, B.R. *Plant Physiol* 2004, 166:525-530.

146. Garbaye, J. *New Phytologist* 1994, 128:197-210.

147. Vivas, A., Vörös, I., Biro, B., Barea, J.M. et al. *Applied Soil Ecology* 2003, 24:177-186.

148. Vivas, A., Vörös, I., Biro, B., Campos, E. et al. *Environmental Pollution* 2003, 126:179-186.

149. Khan, A.G. *In:* The restoration and management of derelict land: modern approaches. Wong, M.H. and Bradshaw, A.D. (eds), World Scientific Publishing, Singapore, 2002, 149-160.

150. Khan, A.G. *In:* Developments in ecosystems. Wong, M.H. (ed.), 2004, 95-114.

151. Bansal, M., Chamola, B.P. and Sarwar, N. *In:* Mycorrhizal Biology. Mukerji, K.G., Chamola, B.P. and Singh, J. (eds), Kluwer Academic/Planum Publishers, New York, 2002, 143-152.

152. del Val, C., Barea, J.M. and Azcon-Aguilar, C. *Applied Soil Ecology* 1999, 11:261-269.

153. Zhu, Y.G., Christie, P. and Laidlaw, A.S. *Chemosphere* 2001, 42:193-199.

154. Jamal, A., Ayub, N., Usman, M. and Khan, A.G. *International Journal of Phytoremediation* 2002, 4:205-221.

155. Sylvia, D.M. and Williams, S.E. *In:* Mycorrhizae in Sustainable Agriculture. Bethlenfalvay, G.J. and Linderman, R.G. (eds), ASA No. 54, Madison, USA, 1992, 101-124.

156. Joner, E., Briones, R. and Leyval, C. *Plant and Soil* 2000, 226:227-234.

157. Tullio, M., Pierandrei, F., Salerno, A. and Rea, E. *Biol Fertil Soils* 2003, 37:211-214.

158. Turnau, K., Kottke, I. and Oberwinkler, F. *New Phytologist* 1993, 123:313-324.

159. Zhou, J.L. *App Microbiol Biotechnol* 1999, 51:686-693.

160. Joner, E.J. and Leyval, C. *New Phytologist* 1997, 135:353-360.

161. Kapoor, A. and Viraraghavan, T. *Biores Technol* 1995, 53:195-206.

162. Driver, J.D., Holben, W.E. and Rilling, M.C. *Soil Biol Biochem* 2005, 37:101-106.

163. Gonzalez-Chavez, M.C., Carrillo-Gonzalez, R., Wright, S.F. and Nichols, K.A. *Environmental Pollution* 2004, 130:317-323.

164. Khan, A.G., Kuek, C., Chaudhry, T.M., Khoo, C.S. et al. *Chemosphere* 2000, 41:197-207.

165. Malcova, R., Vosatka, M. and Gryndler, M. *Applied Soil Ecology* 2003, 23:55-67.

166. Chen, B., Christie, P. and Li, L. *Chemosphere* 2001, 42:185-192.

167. Turnau, K. *Acta Societatis Botanicorum Poloniae* 1998, 67:105-113.

168. Gonzalez-Chavez, C., D'Haen, J., Vangronsveld, J.J. and Dodd, J.C. *Plant and Soil* 2002, 240:287-297.

169. Hildebrandt, U., Kaldorf, M. and Bothe, H. *Journal of Plant Physiology* 1999, 154:709-711.

170. Pawlowska, T.E., Blaszkowski, J. and Rühling, Å. *Mycorrhiza* 1997, 6:499-505.

171. Regvar, M., Groznik, N., Goljevšèek, K. and Gogala, N. *Acta Biol Slov* 2001, 44:27-34.

172. Whitfield, L., Richards, A.J. and Rimmer, D.L. *Mycorrhiza* 2004, 14:55-62.

173. Regvar, M., Vogel-Mikus, K., Kugonic, N., Turk, B. et al. *Environmental Pollution* 2006 144:976-984.

174. Ouziad, F. *Charakterisierung der arbuskulären Mykorrhiza Pilz-Flora an ausgewählten Standorten mit molekularbiologischen Methoden.* Diploma Thesis, University of Cologne, Germany, 1999.

175. Tonin, C., Vandenkoornhuyse, P., Joner, E.J., Straczek, J. et al. *Mycorrhiza* 2001, 10:161-168.

176. Leyval, C., Singh, B.R. and Joner, E.J. *Water Air Soil Pollut* 1995, 84:203-216.

177. Pawlowska, T.E. and Charvat I. *Appl Environ Microbiol* 2004, 70:6643-6649.

178. Del Val, C., Barea, J.M. and Azcon-Aguilar, C. *Appl Environ Microbiol* 1999, 65:718-723.

179. Gildon, A. and Tinker, P.B. *Trans Brit Myc Soc* 1981, 77:648-649.

180. Weissenhorn, I., Leyval, C. and Berthelin, J. *Plant Soil* 1993, 157:247-256.

181. Sambandan, K., Kannan, K. and Raman, N. *J Environ Biol* 1992, 13:159-167.

182. Cavagnaro, T. *Plant and Soil* 2008, 304:315-325.

183. Deram, A., Languereau-Leman, F., Howsam, M., Petit, D. et al. *Soil Biology and Biochemistry* 2008, 40:845-848.

184. Deram, A., Denayer, F.-O., Dubourguier, H.-C., Douay, F. et al. *Science of the Total Environment* 2007, 372:375-381.

185. Audet, P. and Charest, C. *Environmental Pollution* 2007, 147:609-614.

186. Wang, F.Y., Lin, X.G. and Yin, R. *International Journal of Phytoremediation* 2007, 9:345-353.

187. Joner, E.J. and Leyval, C. *Biology and Fertility of Soils* 2001, 33:351-357.

188. Lingua, G., Franchin, C., Todeschini, V., Castiglione, S. et al. *Environ Pollution* 2007, 1-11 doi:10.1016/j.envpol.2007.1007.1012.

189. Lingua, G., Franchin, C., Todeschini, V., Castiglione, S. et al. *Environmental Pollution* 2008, 153:137-147.

190. Bagni, N., Altamura, M.M., Biondi, S., Mengoli, M. et al. *In:* Morphogenesis in Plants. Roubelakis-Angelakis, K.A. and Tran Thanh Van, K. (eds), Plenum Press, New York, 1993, 89-111.

191. Sharma, S.S. and Dietz, K,-J. *J Exp Bot* 2006, 57:711-726.

192. Kozdrój, J., Piotrowska-Seget, Z. and Krupa, P. *Ecotoxicology* 2007, 16:449-456.

193. Sudová, R. and Vosátka, M. *Plant and Soil* 2007, 296:77-83.

194. Janousková, M., Pavlíková, D. and Vosátka, M. *Chemosphere* 2006, 65:1959-1965.

195. Heggo, A., Angle, J.S. and Chaney, R.L. *Soil Biology and Biochemistry* 1990, 22:865-869.

196. Gonzalez-Chavez, C., Harris, P.J., Dodd, J. et al. *New Phytologist* 2002, 155:163-171.

197. Chen, B.D., Li, X.L., Tao, H.Q., Christie, P. et al. *Chemosphere* 2003, 50 839-846.

198. Wang, F.Y., Lin, X.G. and Yin, R. *Pedobiologia* 2007, 51:99-109.

199. Vivas, A., Biro, B., Nemeth, T., Barea, J.M. et al. *Soil Biology and Biochemistry* 2006, 38:2694-2704.

200. Giasson, P., Jaouich, A., Gagné, S. and Moutoglis, P. *Remediation Journal* 2005, 15:113-122.

201. Cooper, K.M. and Tinker, P.B. *New Phytologist* 1978, 81:43-52.

202. Weissenhorn, I., Leyval, C. and Berthelin, J. *Biology and Fertility of Soils* 1995, 19:22-28.

203. Kramer, U. *Current Opinion in Biotechnology* 2005 16:133-141.

204. Chaney, R.L., Li, Y.M., Brown, S.L., Homer, F.A. et al. *In:* Phytoremediation of Contaminated Soil and Water. Terry, N. and Bañuelos, G. (eds), Lewis, Boca Raton, 2000, 129-58.

205. Keller, C., Ludwig, C., Davoli, F. and Wochele, J. *Environ Sci Technol* 2005, 39:3359-3367.

206. Pulford, I.D., Mcgregor, S.D., Duncan, H.J. and Wheeler, C.T. Current topics in plant biochemistry, physiology and molecular biology, Will plants have a role in bioremediation? Interdisciplinary Plant Group, University of Missouri, Columbia, MO, 1995, 49-50.

207. Robinson, B.R.S., Nowack, B., Roulier, S., Menon, M. et al. *FOR Snow Landsc Res* 2006, 80:221-234.

208. Thayalakumaran, T., Robinson, B.H., Vogeler, I, et al. *Plant Soil* 2003, 254:415-423.

209. Tandy, S., Bossart, K., Mueller, R., Ritschel, J. et al. *Environ Sci Technol* 2004, 40:2753-2758.

210. Moreno, F.N., Anaerson, C.W.N., Stewart, R.B., Robinson, B.H. et al. *New Phytologist* 2005, 166:445-454.

211. Roy, S., Labelle, S., Mehta, P. et al. *Plant and Soil* 2005, 272:277-290.

212. Lombi, E., Zhao, F.J., Dunham, S.J. and Mcgrath, S.P. *J Environ Qual* 2001, 30:1919.

213. Angle, J.S., Spiro, M.A., Heggo, A.M., El-Kherbawy, M. et al. *In:* Proceedings of Trace Substances in the Environment Health. 22nd Conference, St. Louis, MO, 1988, 321-336.

214. Lambert, D.H. and Weidensaul, T.C. *Soil Science Society of America Journal* 1991, 55:393-398.
215. Killham, K. and Firestone, M. *Plant and Soil* 1983, 72:39-48.
216. Hetrick, B.A.D., Wilson, G.W.T. and Figge, D.A.H. *Environmental Pollution* 1994, 86:171-179.
217. Mohammad, M.J., Pan, W.L. and Kennedy, A.C. *Soil Science Society of America Journal* 1995, 59:1086-1090.
218. Burke, S.C., Angle, J.S., Chaney, R.L. and Cunningham, S.D. *International Journal of Phytoremediation* 2000, 2:23-29.
219. Bi. Y.L., Li, X.L., Christie, P., Hu, Z.Q. and Wong, M.H. *Chemosphere* 2003, 50:863-869.
220. Piron, E., Accominotti, M. and Domard, A. *Langmuir* 1997, 13:1653-1658.
221. Weng, G.Y., Wang, Z.Q., Wu, L.H., Luo, Y.M. et al. *Soils* 2005, 152-157.
222. Muzzarelli, R., Baldassarre, V., Conto, F., Ferrara, P. et al. *Biomaterials* 1988, 9:247-252.
223. Lou, L.-Q., Shen, Z.-G. and Li, X.-D. *Environmental and Experimental Botany* 2004, 51:111-120.
224. Ernst, W.H.O. *Chemie der Erde* 2005, 65:29-42.
225. Van Nevel, L., Mertens, J., Oorts, K. and Verheyen, K. *Environmental Pollution* 2007, 150:34-40.
226. Dickinson, N.M. *Chemosphere* 2000, 41:259-263.
227. Pulford, I.D. and Watson, C. *Environment International* 2003 29:529-540.
228. Mertens, J., Van Nevel, L., De Schrijver, A., Piesschaert, F. et al. *Environmental Pollution* 2007, 149:173-181.
229. Hildebrandt, U., Regvar, M. and Bothe, H. *Phytochemistry* 2007, 68:139-146.
230. Yoon, J., Cao, X., Zhou, Q. and Ma, L.Q. *Science of the Total Environment* 2006, 368:456-464.
231. Cunningham, S.D., Shann, J.R., Crowley, D.E. and Anderson, T.A. *In:* Phytoremediation of soil and water contaminants. Kruger, E.L., Anderson, T.A. and Coats, J.R. (eds), ACS Symposium series 664, American Chemical Society, Washington, DC, 1997, 2-19.
232. Berti, W.R. and Cunningham, S.D. *In:* Phytoremediation of toxic metals: Using plants to clean-up the environment. Raskin, I. and Ensley, B.D. (eds), Wiley, New York, 2000, 71-88.
233. Schnoor, J.L. *In:* Phytoremediation of toxic metals: Using plants to clean-up the environment. Raskin, I. and Ensley, B.D. (eds), New York, Wiley, 2000, 133-150.
234. Flathman, P.E. and Lanza, G.R. *Journal of Soil Contamination* 1998, 7:415-432.
235. Adriano, D.C., Wenzel, W.W., Vangronsveld, J. and Bolan, N.S. *Geoderma* 2004, 122:121-142.
236. Williamson, A. and Johnson, M.S. *In:* Effect of heavy metal pollution on plants. Lepp, N.W. (ed.), Applied Science Publishers, Barking, Essex, UK, 1981, 185-212.
237. Bolan, N., Adriano, D. and Naidu, R. *Reviews of Environmental Contamination and Toxicology* 2003, 1-44.
238. Berreck, M. and Haselwandter, K. *Mycorrhiza* 2001, 10:275-280.

239. Ouziad, F., Hildebrandt, U., Schmelzer, E. and Bothe, H. *Journal of Plant Physiology* 2005, 162:634-649.
240. Vogel-Mikus, K., Pongrac, P., Kump, P., Necemer, M. et al. *Environmental Pollution* 2006, 139:362-371.
241. Brown, S., Sprenger, M., Maxemchuk, A. and Compton, H. *J Environ Qual* 2005, 34:139-148.
242. Robinson, B.H., Green, S.R., Mills, T.M., Clothier, B.E. et al. *Aust J Soil Res* 2003, 41:599-611.
243. Oudeh, M., Khan, M. and Scullion, J. *Environmental Pollution* 2002, 116:293-300.
244. Denny, H.J. and Ridge, I. *New Phytologist* 1995, 130:251-257.
245. Guo, Y., George, E. and Marschner, H. *Plant and Soil* 1996, 184:195-205.
246. Sharples, J.M., Meharg, A.A., Chambers, S.M. and Cairney, J.W.G. *Plant Physiol* 2000, 124:1323-1327.
247. Sharples, J.M., Meharg, A.A., Chambers, S.M. and Cairney, J.W.G. *Nature* 2000, 404:952-961.
248. Leung, H.M., Ye, Z.H. and Wong, M.H. *Chemosphere* 2007, 66:905-915.
249. Davies, J.F.T., Puryear, J.D., Newton, R.J., Egilla, J.N. et al. *Journal of Plant Physiology* 2001, 158:777-786.
250. El-Kherbawy, M., Angle, J.S., Heggo, A. and Chaney, R.L. *Biology and Fertility of Soils* 1989, 8:61-65.
251. Griffioen, W.A.J. and Ernst, W.H.O. *Agriculture, Ecosystems and Environment* 1990, 29:173-177.
252. Lambert, D.H. and Weidensaul, T.C. *Soil Science Society of America Journal* 1991, 55:393-398.
253. Díaz, G., Azcón-Aguilar, C. and Honrubia, M. *Plant and Soil* 1996, 180:241-249.

12

Molecular Characterization of Genetic Diversity among Arbuscular Mycorrhizal Fungi

M. Velmurugan, P. Hemalatha[1],
C. Harisudan[1] and D. Thangadurai[2]

Horticultural College and Research Institute, Tamil Nadu Agricultural
University, Coimbatore, Tamil Nadu 641003, India

[1]Directorate of Extension Education, Tamil Nadu Agricultural
University, Coimbatore, Tamil Nadu 641003, India

[2]Molecular Breeding Laboratory, Department of Botany, Karnatak
University, Dharwad, Karnataka 580003, India

Introduction

Mycorrhiza is a symbiotic association between plants and fungi during the active period of growth in plants. In 1885, the German forest pathologist A.B. Frank coined the term Mycorrhiza, which literally means fungus root. Fungi of this phylum (Glomeromycota) have a characteristic feature of arbuscules and vesicles which helps in the uptake of phosphorus and micronutrients from the soil. Additionally it confers the resistance against plant pathogens.

The paleobiological and molecular evidences confine that arbuscular mycorrhiza (AM) might have originated at least 460 million years ago. It is ubiquitous in nature and 80% of AM are found in terrestrial ecosystem [1] and might have existed in early ancestors of extant land plants. These positive symbiotic associations may have facilitated the development of land plants. AM fungi represent a large underground interconnecting mycelial network [2].

In the rhizosphere, enormous variations are noticed in AM fungi. These variations can effectively be studied with the help of molecular characterization tools viz., RAPD, AFLP etc. Genetic diversity is level of biodiversity and refers to the total genetic characteristics or genetic makeup of a species otherwise it is called as any variation in the nucleotides, genes, chromosomes, or whole genomes of organisms. It is primarily represented by the variations in the nucleotides that form the DNA within the cells of the organism.

Genetic Diversity in Arbuscular Mycorrhizal Fungi

The theories and hypothesis pertaining to the genetic diversity can be studied in the population genetics. It plays a vital role in survival and adaptability of a species. Due to the environment changes, slight gene variations are necessary to adapt or to acclimatize and survive in that specific area. A species that has a large degree of genetic diversity among its population will have more variations. Species that have very little genetic variation are at a great risk. With very little gene variation within the species, healthy reproduction becomes increasingly difficult and offspring often deal with problems similar to inbreeding.

Any alteration of the AM fungal diversity will modify the rhizospheric potentialities and ultimately the soil fertility will be affected. AM spores seem to contain high numbers of nuclei [3] and intrasporal variations between nuclei have been suggested for *Gigaspora margarita* [4] possibly because of hyphal anastomosis, which do not entail meiosis [5, 6]. The intraspecific diversity between *Glomus claroideum* and *Glomus* DAOM 225952 revealed that the range was high, from 22 to 33 different electrophoretic types for *G. claroideum* and 15-27 for *Glomus* DAOM 225952 depending on the population. The differences in hyphal architecture and growth patterns of Glomeraceae and Gigasporaceae are given in Table 1.

The distribution of phylogenetic groups of AM fungi belonging to a clade of Glomus species was studied in five plant species (*Hypochoeris radicata, H. pilosella, Thymus serpyllum, Artemisia campestris* and *Armeria maritima*) from a coastal grassland in Denmark. The results showed that the dominant *Glomus* species were able to colonize all the studied plant species, supporting the view that the AM fungi represent a large underground interconnecting mycelial network. Since arbuscular mycorrhizal fungi lack a tractable genetic system,

Table 1. Differences in hyphal architecture and growth
patterns of Glomeraceae and Gigasporaceae [7]

Gigasporaceae	*Glomeraceae*
Gigasporaceae can colonize only roots from germinating spores and do not form cross-links among hyphae.	Glomeraceae are able to colonize roots from mycelium fragments or colonized root pieces.
When the hyphae are injured it has an inherent potentiality of repairing of its own mechanisms.	They repair injured hyphae by forming a network of cross-links instead of repairing the main hyphal axis.

vegetative compatibility tests may represent an easy assay for the detection of genetically different mycelia and an additional powerful tool for investigating the population structure and genetics of these obligate symbionts [8].

Classification of the Order Glomales

Order: Glomales Morton & Benny
    Suborder: Glomineae Morton & Benny
        Family: Glomaceae Pirozynski & Dalpe
            Genus: *Glomus* Tulasne & Tulasne
            Genus: *Sclerocystis* (Berkeley & Broome) Almeida & Schenck
        Family: Acaulosporaceae Morton & Benny
            Genus: *Acaulospora* (Gerdemann & Trappe) Berch
            Genus: *Entrophosphora* Ames & Schneider
    Suborder: Gigasporineae Morton & Benny
        Family: Gigasporaceae Morton & Benny
            Genus: *Gigaspora* (Gerdemann & Trappe) Walker & Sanders
            Genus: *Scutellospora* Walker & Sanders

The earliest evidences for AM symbiosis were noticed in fossil arbuscules found in specimens of land plant *Algaophyton*. Fossil spores and hyphae resembling those of present glomalean fungi had been detected in some plant materials. It has been observed that Ordovician spores and hyphae are similar to *Glomus* type 60. These findings confirm that glomalean fungi were present at a time when the land flora comprised plants of the bryophyta class, lending strong support to the notion that AM fungi were instrumental in the succession of early land plants [9, 17, 28].

Molecular Characterization of Arbuscular Mycorrhizal Fungi

The first molecular phylogeny of AM fungi was reported by [10] using ribosomal small subunit (SSU) sequences. Understanding the molecular and functional approaches of AM is difficult because of unusually high within-individual genetic diversity [11]. The altered fungal diversity affects the soil fertility and affects the growth rates. Currently Mycorrhizae is noticed in 92% of plant families (80% of species) with vesicular arbuscular mycorrhizae being the ancestral and predominant form [12] and indeed the most prevalent symbiotic association found in all plant kingdom [13].

The results of Amplified fragment length polymorphism (AFLP) [14] revealed the high genetic variation in AM fungi at both interspecific and intraspecific levels. The molecular diversity of glomalean fungi was studied by analysis of *Glomus* specific sequences amplified from field harvested peas and used Single-stranded conformation polymorphism (SSCP) method for screening. The results revealed that the root derived sequences were aligned with sequences from spores of 17 cultured *Glomus* isolates; five of these isolated from the same field from where the peas were harvested [15].

Glomeromycotan fungi are obligate symbionts and the progress of obtaining new marker genes has been delayed because their spores usually contain unusual number of other microorganisms, including fungi of other phyla [16]. Due to the problems outlined above there is currently no molecular species concept for glomeromycotan fungi. Nevertheless molecular markers have proven to be highly useful to characterize the diversity of AM fungi in the field conditions.

In 1974, Gerdemann and Trappe placed arbuscular mycorrhizal fungi in four different genera in the order Endogonales (*Glomus, Sclerocystis, Gigaspora, Acaulospora*). Later in 1990, 'Glomales' was placed in the Zygomycota, comprising six genera. Since then evidence has accumulated supporting the view that arbuscular mycorrhizal fungi are distinct from other Zygomycota [17]. The results of rDNA phylogeny revealed that the AM fungi are the sister group of Asco- and Basidiomycota and not monophyletic with any part of the Zygomycota. Hence, the 'Glomales' was raised to the rank of a phylum Glomeromycota [18].

Ten genera of the Glomeromycota currently are distinguished as follows [19]:

(i) *Glomus* is the largest genus in the phylum consisting of more than 70 morphospecies. The glomoid mode of spore formation is symplesiomorphic and occurs in several distinct lineages, namely *Glomus, Paraglomus, Archaeospora, Pacispora, Diversispora* and *Geosiphon*. Certain genera were separated from *Glomus* based on molecular phylogenetic characteristics. Molecular analysis showed that it is nested within a clade of other well characterized *Glomus* species, hence this species also was transferred to *Glomus* [20].

(ii, iii) The family Gigasporaceae contains two closely related species viz., *Gigaspora* and *Scutellospora*. Spores are formed on a bulbous sporogeneous cell and germinate through a newly formed opening in the spore wall. These two genera do not form vesicles within roots and the extraradical mycelium bears auxiliary cells of unknown function. In contrast to *Gigaspora*, species of *Scutellospora* possess flexible inner spore walls with permanent mature spores. *Gigaspora* germinates after a papillate layer has formed on the inner of the spore wall.

(iv, v) The diagnostic feature of the family Acaulosporaceae is the formation of spores next to a 'sporiferous saccule'. During maturation this saccule collapses and disappears. Another feature is that it contains flexible inner walls or germinal walls. During spore germination a 'germination orb' is produced on the inner walls, which is a membraneous structure that is instrumental in penetrating the outer spore wall. It is produced laterally in *Acaulospora* and formed within the subtending hypha in *Entrophospora*. The phylogenetic position of *E. infrequens*, the type species of the genus, is unclear because rDNA sequences from several unrelated glomeromycotan lineages were reported to occur within its spores [21].

(vi) The recently established genus for AM fungal species was *Pacispora* which forms spores in the same way *Glomus* typically but it is having flexible inner walls and a germination orb [22].

(vii) Another clade of *Glomus* species, referred to as *Glomus* group C, is more closely related to the Acaulosporaceae based on rDNA phylogenies [23]. Among them, one species has been described in a new genus *Diversispora* as *D. spurca*, mainly based on ribosomal sequence signatures.

(viii) Those AM fungi which are deeply divergent within the phylum are established under the genus *Archaeospora* [20, 24], *A. leptoticha* and *A. gerdemannii* are dimorphic, producing both glomoid and acaulosporoid spores. Most of the isolates produce both types of spores at the same time. The acaulosporoid spores of *Archaeospora* show distinct characters from *Acaulospora*.

(ix) The only member of the phylum that is known to engage in a different type of symbiosis is *Geosiphon pyriformis* [25]. The exact phylogenetic relationship of *Geosiphon* is relative to *Paraglomus* and *Archaeospora*.

(x) *Paraglomus* is another species that forms small, hyaline spores that do not show light microscopic characters but distinguish from *Glomus* species. The results of rDNA phylogenies clearly showed that *Paraglomus* is not related to other *Glomus* species and is basal to the phylum [20]. The basal position of *Archaeospora* and *Paraglomus* is supported peculiarly by unique fatty acids which is unnoticed in other glomeromycotan fungi [26].

Based on the results of green house experiments, it has been observed that genetically different *Glomus intraradices* isolates from one AMF population significantly alter plant growth in an axenic system. This confirms that genetic variability in AMF populations could affect host-plant relationships [27].

Experimentation of AM fungi genome is diffident due to the fact that these fungi possess large genomes compared to other Zygomycetes. Molecular methods have been particularly successful for studying rDNA sequences from AMF [25-27]. In addition to this, researchers have reported that individual spores of AM fungi are multinucleate, and show a high level of genetic diversity. Ribosomal-based DNA sequence analysis has revealed genetic variation both within and between AM fungi species. Molecular characterization involving identification of genetic diversity demands sufficient quantities of genomic DNA. However, it could not be cultured due its obligate symbiotic nature. The details of some of the PCR-based molecular studies on AMF are given in Table 2.

Table 2. Details of some PCR-based molecular studies on AMF [28]

Amplified region	Molecular marker	Primer(s)	Target organism	Reference
SSU rDNA	PCR	VANSI	Glomales	[29]
Genomic DNA	PCR-RAPD	OPA-02 and OPA-04	Glomus versiforms, G. mosseae	[30]
		OPA-18 and P124	G. caledonium, Acaulospora laevis	
		OPA-18 and P124	Gigaspora margarita, Scutellospora gregaria	
SSU rDNA	PCR	VANSI and NS21	G. intraradices	[31]
Genomic DNA	Competitive PCR	PO and M3	G. mosseae	[32]
ITS	PCR-RFLP	ITS1 and ITS4	Glomus sp., Scutellospora sp., Gigaspora sp.	[33]
SSU 1492	PCR	NS71 and SSU1492'	Gigaspora sp.	[34]
Partial rDNA	PCR-Partial	SS38 and VANSI	Roots and spores of AM	[35]
		VANSI	Scutellospora and Glomus	
		VAGIGA	Gigasporaceae	
ITS1 and ITS2	PCR	ITS1 and ITS2	G. margarita	[36]
ITS	PCR	ITS1 and ITS4	G. mosseae and Gigaspora margarita	[37]
SSU rDNA	PCR-RFLP	LR1 and FLR2	Subgroups of Glomales	[38]
	PCR-nested	FLR2-5.23 and FLR2-8.23	G. mosseae, G. intraradices	
		LR1-23.46	G. roseae	
28S rDNA	PCR-SSCPs	LSU-Primers	Glomus sp.	[39]
SSU rDNA	PCR	NS31 and AM1	Glomus sp.	[40]
SSU rDNA	PCR-SSCP	VANSI	Subgroups of Glomaes	[41]
	Nested PCR	ITS, AM1	Glomus sp.	
ITS	PCR-RFLP	ITS1 and ITS4	G. mosseae	[42]
ITS	Nested PCR-SSCP	Eukaryotic universal primer	Glomus sp.	[43]
		Glomus-specific ITS primer		
ITS	Nested PCR	ITS5 and ITS4	Glomeromycota (except Archaeosporaceae)	[44]
ITS	PCR	SSU-Glom/LUS-Glom 1	Major groups within Glomeromycota	
		ITS5 and ITS		

Constraints in the Use of Molecular Techniques on AM Fungi

The molecular characterization of AM fungi gave promising results in demarcating one species from another. Certain problems are unique to this group and specifically noticed in other organisms also. Since AM fungi are deeply residing in the plant roots, it could be possible to extract the ample quantity of DNA which is essential for molecular characterization. Certain specific primers are required to characterize them. Without specific primers other pathogenic and saprophytic microorganism will also be detected. Designing of primer for glomalean fungi has proven to be difficult. One piece of root can be colonized by different AM fungi and multiple component colonizers have to be separated by cloning the PCR products [45, 46]. rDNA of AM fungi spores are highly polymorphic in nature when compared to other fungi. Single genetic locus (e.g. rDNA) does not provide clear differentiation of genetic variation between intra-species and inter species. This problem was addressed by establishing species concepts based on gene geneology.

Conclusion and Future Perspectives

AM fungi have its potentiality of mobilization of phosphorus and make them available to the plants. It is present in the soil and colonizes the plant roots by symbiotic association. Therefore it is essential to understand the mechanism and functions of symbiosis in mycorrhizae. Even though with certain constraints, it is mandatory to know the genetic diversity and molecular characterization for in sight level of research works. Therefore, research work needs to be done to elucidate how the heterogeneity of nuclei is maintained in the mycelium. Based on the finding of present and past it is inferred that it is making inroads into the problematic and unapproachable areas of AM fungi. Many unforeseen details has to be holistically studied and reported. The development of glomalean-specific primer and AM transformation techniques may help to link gene expression studies with structural aspects of symbiosis.

References

1. Smith, S.E. and Read, D.J. Mycorrhizal Symbiosis, 2nd ed., Academic Press, London, 1997.
2. Stukenbrock, E.H. and Rosendahl, S. *Mycorrhiza* 2005, 15(7):0940-6360.
3. Biancotto, V. and Bonfante, P. *Protoplasma* 1993, 176:100-105.
4. Lanfranco, L., Delpero, M. and Bonfante, P. *Mol Ecol* 1999, 8:37-45.
5. Giovannetti, M., Azzolini, D. and Citernesi, A.S. *Appl Env Microbiol* 1999, 65:5571-5575.

6. Hosny, M., Hijri, M., Passerieux, E. et al. *Gene* 1999, 226:61-71.
7. Hijri, M., Niculita, H. and Sanders, I.R. *Heredity* 2001, 87:243-253.
8. Giovannetti, M., Sbrana, C., Strani, P. et al. *Appl Env Microbiol* 2003, 16:616-624.
9. Harrier, L.A. *J Exp Bot* 2001, 52:469-478.
10. Simon, L., Bousquet, J., Levesque, C. et al. *Nature* 1993, 363:67-69.
11. Sanders, I.R. *Am. Nat.* 2002, 160:128-141.
12. Wang, B. and Qiu, Y.L. *Mycorrhizahello* 2006, 16(5):299-363.
13. Harrison, M.J. *Annu Rev Microbiol.* 2005, 59:19-42.
14. Rosendahl, S. and Taylor, J.W. *Mol Ecol* 1997, 6:821-829.
15. Kjoller, R. and Rosendahl, S. *Mycological Research* 2001, 105(9):1027-1032.
16. Hijri, I., Sykorova, Z., Oehl, F. et al. *Mol Ecol* 2006, 15:2277-2289.
17. Morton, J.B. and Benny, G.L. *Mycotaxon* 1990, 37:471-491.
18. Schubler, A. *Plant Soil* 2002, 244:75-83.
19. Redecker, D. and Raab, P. *Mycologia* 2006, 98(6):885-895.
20. Redecker, D., Kodner, R. and Graham, L.E. *Science* 2000, 289:1920-1921.
21. Rodriguez, A., Dougall, T., Dodd, J.C. et al. *New Phytol* 2001, 152:159-167.
22. Oehl, F. and Sieverding, E. *J App Bot* 2004, 78:72-82.
23. Schwarzott, D., Walker, C. and Schubler, A. *Mol Phylogen Evol* 2001, 21:190-197.
24. Sawaki, H., Sugawara, K. and Saito, M. *Mycoscience* 1998, 39:477-480.
25. Schubler, A., Mollenhauer, D., Schnepf, E. et al. *Bot Acta* 1994, 107:36-45.
26. Graham, J.H., Hodge, N.C. and Morton, J.B. *Appl Environ Microbiol* 1995, 61:58-64.
27. Alexander, M., Koch, Croll, D. et al. *Ecology Letters* 2004, 9(2):103-110.
28. Reddy, S.R., Pindi, P.K. and Reddy, S.M. *Current Science* 2005, 89(10):1699-1709.
29. Simon, L., Levesque, R.C. and Lalonde, M. *Appl Environ. Microbiol.* 1992, 59:4211-4215.
30. Wyss, P. and Bonfante, P. *Mycol. Res* 1993, 97:1351-1357.
31. Di Bonito, R., Elliott, M.L. and Des Jardin, E.A. *Appl. Environ. Microbiol.* 1995, 61:2809-2810.
32. Edwards, S.G., Fitter, A.H. and Young, J.P.W. *Mycol. Res.* 1997, 10:1440-1444.
33. Redecker, D., Thierfelder, H., Walker, C. et al. *Appl. Environ. Microbiol.* 1997, 63:1756-1761.
34. Bago, B., Bentivenga, S.P., Brenec, V. et al. *New Phytol.* 1998, 139:581-588.
35. Clapp, J.P., Fitter, A.H. and Young, J.P. *Mol. Ecol.* 1999, 8:915-921.
36. Marianne, L., Van Lanfranco, L., Longato, S. et al. *Mycol. Res.* 2003, 103:955-960.
37. Antoniolli, Z.I., Schachtman, D.P., Ophel, K.K. et al. *Mycol. Res.* 2000, 104:708-715.
38. Jacquot, E., Van Tuinen, D., Gianinazzi, S. et al. *Plant Soil* 2000, 226:179-188.
39. Kjoller, R. and Rosendahl, S. *Plant Soil*, 2000, 226:189-196.
40. Daniell, T.J., Husband, R., Fitter, A.H. et al. *Ecol.* 2001, 36:203-209.
41. Redecker, D. *Plant Soil* 2002, 244:67-73.
42. Giovannetti, M., Sbrana, C., Strani, P. et al. *Appl. Environ. Microbiol.* 2003, 69:615-624.
43. Kjoller, R. and Rosendahl, S. *Mycol. Res.* 2003, 105:1027-1032.
44. Renker, C., Heinrichs, J., Kaldorf, M. et al. *Mycorrhiza*, 2003, 13:191-198.
45. Clapp, J.P., Young, J.P.W., Merryweather, J.W. et al. *New Phytol.* 1995, 130:259-265.
46. Van Tuinen, D., Jacquot, E., Zhao, B. et al. *Mol. Ecol.* 1998, 7:879-887.

13

Molecular Tools for Biodiversity and Phylogenetic Studies in Mycorrhizas: The Use of Primers to Detect Arbuscular Mycorrhizal Fungi

Fernanda Covacevich

Universidade Federal Rural do Río de Janeiro, Dep. de Solos, BR 465 km 7, CEP 23890-970 Seropédica, RJ, Brasil; Inter-American Institute for Global Change Research (IAI) CRN II/14 Supported by the US National Science Foundation

Introduction

The association between arbuscular mycorrhizal fungi (AMF) and root plants occurs over a broad ecological range, from aquatic to desert environments [1]. The global distribution, beneficial effects on plant growth and ecological importance of AMF has been well documented. However, knowledge of their community structure is scarce. In recent years, more information has been reported regarding the functional role of AMF in ecosystems [2]. Studies under mesocosm, field plot and natural conditions suggest that belowground diversity of AMF may influence vascular plant community structure and composition [3, 4]. In order to understand factors structuring plant communities, more information is needed about the natural distribution patterns of AMF. It is thus necessary to identify the fungi associated with natural ecosystems and agroecosystems. Only about 150 AMF species have been formally described [5]. Establishment of phylogenetic relations, identification and classifications are based on morphological features of the asexually produced propagules (spores, sporocarps). In the absence of spores, the intraradical structures allow identification to the family level at the most [6]. However, in most cases the identification of AMF is generally difficult or almost impossible [7].

Morphological characters of spores may leave many species unresolved. However, even when they can be identified, basing our understanding of AM fungal communities on spores in the soil is like basing studies of plant communities only on the soil seed bank available. Another drawback of conventional approaches is the inability to isolate fungi, because they form obligate symbiotic associations with the partner [8]. All these limitations often lead to underestimate fungal diversity, population number and species richness. Molecular techniques based on DNA analysis seem to offer a wide range of advantages. Furthermore, the combination of molecular biology methods may be the most promising way to monitor community structure and biodiversity of AMF in the field.

Molecular approaches primarily rely upon utilization of genetic variation. However, molecular studies have shown that spore populations in the soil do not always reflect the AM fungal communities present in roots [9, 10]. One methodological advance in the study of mycorrhiza has been the application of the polymerase chain reaction (PCR). This has led to the development of techniques that are not limited by the culturability of fungi [11-15]. Ribosomal-based DNA (rDNA) sequence analysis has revealed genetic variation both within and between AMF species. Advances in the phylogeny of the Glomeromycota based on rDNA sequences have demonstrated that some highly divergent taxa are not distinguishable by their morphological characteristics [16]. In all eukaryotic organisms there are multiple copies of the rDNA per cell; the ribosomal genes possess highly conserved sectors which facilitate the design of primers to hybridize successfully in the region to be amplified. In addition, there are variable regions that allow the differentiation of taxa at different levels. Ribosomal genes are multicopy genes tandemly organized in the genome, separated from each other by an Inter Genic Spacer (IGS). Ribosomal genes are comprised of three subunits of coding regions (18S[SSU], 5.8S and 25S[LSU]), separated from each other by an Inter Non Transcribed region (ITS). There are conserved areas which (1) can be used as universal primer sites, and (2) aid in alignment of sequences prior to phylogenetic analyses [17-20]. The 18S rDNA is the most useful region for phylogenetic and biodiversity studies. However, it has some limitations because it is the most variable region of the nuclear ribosomal genes, and show high intraspecific variation [17, 21]. Schüßler et al. [22] reported that the within-isolate variation of the 18S rDNA is relatively small; thus, that AM fungal phylogeny could be based on this gene. They established the relationships among various AMF, and between AMF and other fungi by using molecular data. The 5' end of the 25S ribosomal subunit harbours two informative polymorphic domains (D1 and D2). The polymorphism observed in these domains between and within taxa allows identification of specific nucleotidic sequences. These sequences can be used to design primers with different levels of specificity or discrimination [23].

The ITS and IGS are variable regions which mutate more frequently than the three conserved coding subunit regions. This generally makes ITS and IGS more informative for analyses of closely related genomes. Coding regions of the small and large ribosomal subunits are considered to be more useful for understanding more distant relationships at the species/order level. The ITS and IGS regions evolve sequence differences between different populations of the same species, or within single spores in the case of the Glomales. Jansa et al. [21] confirmed high levels of ITS variation in Glomeromycota species/phylotypes. Furthermore, Redecker [24] mentioned that the topology of the 5.8S neighbour allows the separation of all major groups of Glomeromycota. However, Renker et al. [25] found that the ITS region is not adequate to reconstruct the phylum Glomeromycota. In addition, the whole ITS gives access to fine population analyses, due to high levels of variability within ITS1 and ITS2.

The first molecular study on AMF was made by Simon et al. [26], and the polymorphic nature of the ribosomal DNA in AMF was first described by Sanders et al. [17]. Subsequently, more studies reported the development of primers with improved success in specific amplification of the glomalean rDNA. At the same time, a large number of publications on AM fungal molecular ecology appeared. The analyses of rDNA gene cluster aimed to identify AMF colonizing plant roots, and it made possible to study the diversity of AMF *in planta* with a high degree of precision and reproducibility [8, 9, 18, 22, 23, 27-31]. However, many of the AMF sequences collected from field samples do not match sequences from known, pot-cultured AMF. Thus, currently, sequences are partitioned into groups based on their similarities, and cannot be assigned to a particular species. Much information about rDNA sequences with the species specific name has been accumulated in a gene bank NCBI (http://www.ncbi.nlm.nih.gov). It was also possible to confirm the biological species with their specific primers. However, there is no information about the primers commonly used for phylogenetic and biodiversity studies of arbuscular mycorrhizal relationships, and in some cases information seems to be repeated. This chapter collects information from spores to mycorrhizal roots about the most useful primers for phylogenetic and biodiversity studies of AMF.

Molecular Techniques Commonly Used for Mycorrhizal Phylogenetic and Biodiversity Studies

Direct evidence shows that individual spores of AMF are multinucleated; that is, one AMF species is heterokaryotic containing populations of genetically different nuclei [19]. Thus, molecular studies for taxonomy purposes are preferably conducted on a single spore. For biodiversity studies, however, DNA extraction and PCR amplification from AM colonized roots are also common. Sequencing PCR products derived from mycorrhizal roots and spores will help developing new insights into 'species' diversity. Furthermore, a diverse range of molecular techniques without the need for sequencing have been applied to the study of biodiversity of AMF. They include (1) restriction fragment length polymorphism (PCR–RFLP) [18, 20, 28]; (2) terminal (t)-RFLP [20, 32]; (3) single stranded conformation polymorphism (SSCP) [21, 29]; (4) denaturing gradient gel electrophoresis (DGGE) [33-36]; and (5) minisatellites, among others.

Sanders et al. [17] characterized biodiversity of AMF present in natural populations from a unique spore by analyzing the RFLP pattern. Vandenkoornhuyse et al. [20] reported the diversity of the AM fungal community composition in the roots of *Agrostis capillaries* and *Trifolium repens*, that co-occurred in the same grassland ecosystem, by using the same technique. The study demonstrated that 19 of these phylotypes belonged to the Glomaceae, three to the Acaulosporaceae and two to the Gigasporaceae. However, in some cases, phylogenetic analysis showed that all the obtained clones belonging to a genus could not be identified at the species level. In the last years, T-RFLP analysis is becoming increasingly popular for examining AM fungal communities in environmental samples [32, 37, 38]. These methods involve end-labelling PCR amplicons with fluorescent molecules attached to the 5'-end of one or both PCR primers. Sequence heterogeneity between rDNA of different species or phylogenetic groups results in different terminal restriction fragment (T-RF) sizes, when PCR amplicons are digested with select restriction enzymes. After electrophoretic separation of the resulting fragments on polyacrylamide gel or capillary DNA sequencers, T-RF size distributions are analyzed by laser excitation and visualization of the fluorine. T-RF size distributions can be compared between samples to yield measures of community similarity which can be analyzed using multivariate statistical methods.

The PCR-SSCP is a simple procedure where denatured PCR products are subject to electrophoresis through a non-denaturing polyacrylamide gel. The distinguishing patterns obtained with PCR-SSCP are sequence-dependent and utilize minor nucleotide differences across several hundred base sequences. Each PCR product with a different sequence, therefore, will theoretically be represented by two bands which correspond to the two strands of the amplified molecule. SSCP was shown to detect single base changes in either 99% or 89% of PCR products having between 100-300 or 300-450 base pairs, respectively [21, 29]. SSCP allows use of variation levels that are seldom available to other techniques. In practice, however, sequence differences between species in variable regions such as the ITS, are frequently represented by more than a single base change and so separation does not usually rely upon such high levels of sensitivity.

The DGGE approach has been applied in microbial ecology as a sensitive and rapid technique for profiling microbial communities. Muyzer et al. [39] was the first to use PCR-DGGE to profile microbial communities. Primers for DGGE studies are the same than those from other studies (cloning, RFLP, SSCP), plus a GC clamp which is necessary to help the melting behaviour of the amplicons. Separation of amplified DNA by PCR in DGGE is based on differences in sequence composition that affect its melting behaviour. This causes a decrease in the electrophoretic mobility of a partially melted DNA molecule in a polyacrylamide gel, which contains a linearly increasing gradient of DNA denaturants. The first use of this technique for fungal community analysis was by Kowalchuk et al. [33]. Since then, PCR-DGGE has proven to be a powerful technique for the culture-independent detection and characterization of fungal populations in plant material and soil without the cloning process [14, 34, 40]. PCR-DGGE is complimentary to cloning strategies for fungal community studies, by tentatively identifying cloned 18S rDNA fragments by comparison to community DGGE banding patterns [14]. Vainio and Hantula [40] showed that DGGE detected more fungal species from environmental samples than culturing techniques. Kowalchuk et al. [34] applied PCR-DGGE to study the AMF community structure at the field. They noted discrepancies observed between the AM fungal-like groups detected in spore populations versus direct 18S rDNA analysis of root material by DGGE. Suggestions that spore inspection alone may poorly represent actual AM fungal population structure were thus corroborated.

Evidence of repeated DNA sequences has been reported elsewhere in genomes of AMF [41, 42]. The possibility of using a tandemly repeated DNA sequence as a diagnostic probe for AMF detection in colonized roots has been demonstrated previously [42]. In this way, minisatellites could be used when the DNA sequences of the ITS region cannot serve as molecular markers for the identification of some AMF.

Primers Commonly Used for Arbuscular Mycorrhizal Phylogenetic and Biodiversity Studies

Most AMF phylogenetic and biodiversity studies have utilized the universal eukaryotic primers designed in the 1990s [43-46] (Table 1). Most of the times, these studies have used specific mycorrhizal primers to amplify the 18S rDNA of AMF. Then, general fungal primers have been designed to amplify all fungal 18S rDNA; this is based on the fact that sequence must be representative of all phyla of fungi. Furthermore, universal eukaryotic primers have been designed to amplify fragments of the ITS regions or the 25S rDNA (Tables 2 and 3). Also, they have been successfully used in combination with AMF specific primers. In some cases, the PCR approach is based on the use of degenerative primers. Furthermore, in some cases a unique PCR reaction does not produce DNA amplification, and nested PCR amplification must be performed. The nested PCR approach involves two sets of primers in two steps of amplification; this is commonly used in AMF research to overcome PCR inhibition and to increase sensitivity for rare DNA templates. The nested PCR approach involves, in general, (1) initial amplification with universal or general fungal primers with target fragments of the whole fungal community, and (2) subsequent amplification on the diluted products from the first PCR with taxon-discriminating primers.

Table 1. Universal eukaryotes and general fungal primers used to amplify fragments of the 18S rDNA of arbuscular mycorrhizal fungi

Primer name	Primer sequence	Target group	rDNA region	Reference
NS1	5'-GTAGTCATATGCTTGTCTC-3'	Universal eukaryotes	Starting of 18S	[43]
NS2	5'-GGCTGCTGGCACCAGACTTGC- 3'	Universal eukaryotes	18S	[43]
NS3	5'-GCAAGTCTGGTGCCAGCAGCC- 3'	Universal eukaryotes	18S	[43]
NS4	5'-CTTCCGTCAATTCCTTTAAG-3'	Universal eukaryotes	18S Middle of	[43]
NS5	5'-AACTTAAAGGAATTGACGGAAG-3'	Universal eukaryotes	18S	[43]
NS6	5'-GCATCACAGACCTGTTATTGCCTC-3'	Universal eukaryotes	18S	[43]
NS7	5'- GAGGCAATAACTGGTCTGTGATGC-3'	Universal eukaryotes	18S (V9)	[43]
NS8	5'-TCCGCAGGTTCACCTACGGA-3'	Universal eukaryotes	18S	[43]
SS38	5'-GTCGACTCCTGCCAGTAGTCATATGCTT- 3'	Universal eukaryotes	18S	[44]
SS1492	5'-GCGGCCGCTACGGMWACCTTGTTACGACTT-3'	Universal eukaryotes	18S	[44]
NS21	5'-AATATACGCTATTGGAGCTGG-3'	Universal eukaryotes	18S	[45]
NS31	5'-TTGGAGGGCAAGTCTGGTGCC-3'	Universal eukaryotes	18S (V3-V4)	[45]
NS41	5'-CCCGTGTTGAGTCAAATTA-3'	Universal eukaryotes	18S	[45]
NS51	5'-GGGGGAGTATGGTCGCAAGGC-3'	Universal eukaryotes	18S	[45]
NS61	5'-CAGTGTAGCGCGCGTGCGGC-3'	Universal eukaryotes	18S	[45]
NS20	5'-CGTCCCTATTAATCATTACG-3'	Universal eukaryotes	18S	[62]
FR1	5'-AICCATTCAATCGGTAIT-3'	General fungal	18S	[40]
GeoA1	5'- GGTTGATCCTGCCAGTAGTC-3'	General fungal	18S	[46]
GeoA2	5'-CCAGTAGTCATATGCTTGTCTC-3'	General fungal (Represents NS1 elongated by 3 bp at the 5' end)	18S	[46]
Geo10	5'-ACCTTGTTACGACTTTTACTTC-3'	General fungal	18S	[46]
Geo11	5'-ACCTTGTTACGACTTTTACTTCC-3'	General fungal	18S	[46]
GeoNS1	5'-ATGGCTCATTAAATCAGTTAT-3'	General fungal	18S	[46]
ART4	5'-TCCGCAGGTTCACCTACGG-3'	General fungal	18S	[46]
F1Ra	5'-CTTTTACTTCCTCTAAATGACC-3'	General fungal	End of 18S	[35]

Table 2. Universal eukaryotes and general fungal primers used to amplify fragments of the Internal transcribed spacer (ITS) region of arbuscular mycorrhizal fungi

Primer name	Primer sequence	Target group	rDNA region	Reference
ITS1	5′- TCCGTAGGTGAACCTGCGG-3′	Universal eukaryotes	Starting of ITS	[43]
ITS2	5′-GCTGCGTTCTTCATCGATGC-3′	Fungi and Basidiomycetes	ITS (similar to 5.8S)	[43]
ITS3	5′-GCATCGATGAAGAACGCAGC-3′	Universal eukaryotes	ITS	[43]
ITS4	5′-TCCTCCGCTTATTGATATGC-3′	Universal eukaryotes	End of ITS (ITS4)	[43]
ITS5	5′-GGAAGTAAAAGTCGTAACAA GG-3′	Universal eukaryotes	ITS	[43]
ITS1-F	5′- CTTGGTCATTTAGAGGAAGTAA-3′	Fungi and Basidiomycetes	ITS	[63]
ITS4 B	5′-CAGGAGACTTGTACACGGTCCAG-3′	Basidiomycetes	ITS	[63]
5.8s	5′-CGCTGCGTTCTTCATCG-3′	General fungal	5.8S (ITS3)	[64]
5.8Sr	5′- TCGATGAAGAACGCA GC-3′	General fungal	5.8S (ITS3)	[64]
ITS26	5′-ATATGCTTAAGTTCAGCGGGT-3′	Universal eukaryotes	ITS	[65]
ITS1-26	5′- TCCGTAGGTGAACCTGCGGGAAGGATC-3′	Universal eukaryotes	ITS	[55]

Table 3. Universal eukaryotes and general fungal primers used to amplify fragments of the 25S rDNA region of arbuscular mycorrhizal fungi

Primer name	Primer sequence	Target group	rDNA region	Reference
LSU 0061	5′-AGCATATCAATAAGCGGAGGA-3′	Universal eukaryotes	5′ end of the 25S (corresponds to LR1)	[66]
LSU 0599	5′- TGGTCCGTGTTTCAAGACG-3′	Universal eukaryotes	25S (corresponds to NDL22)	[66]
LR1	5′-GCATATCAATAAGCGGAGGA-3′	Universal eukaryotes	D1 region of the 25S rDNA	[57]
NDL22	5′- TGGTCCGTGTTTCAAGACG-3′	Universal eukaryotes	D2 region of the 25S rDNA	[57]
FLR2	5′-GTCGTTTAAAGCCATTACGTC -3′	General fungal	25S	[58]

Simon et al. [47] designed primers to identify AMF (Table 4) in colonized roots by PCR fragment amplification of the 18S rDNA combined with the SSCP analysis. The VALETC, VAGLO, VAACAU, and VAGIGA primers were designed to discriminate among four distinct groups of endomycorrhizal species. Furthermore they designed the VANS22 and VANS32, which were able to amplify a 150-bp informative fragment from any endomycorrhizal fungi. These primers were not designed specifically for Glomales, and Simon et al. [47] mentioned the possibility that they could also be useful for other fungi or eukaryotes. Later, Simon and Lalonde [45] designed and patented the VANS1 primer, which amplified part of the 18S rDNA from AMF (*Glomus intraradices* and *Gigaspora margarita*) directly from colonized roots. However, Simon [48] concluded that primers pairs VANS22/VANS32 and NS71/SSU1492 can only detect AMF genus differences. Furthermore, he reported that those primers are not AMF specific and samples must be treated with a nested PCR: amplified firstly by the primers VANS1 (Glomalean specific) and VANS22, and then the amplicons amplified by the VANS22/VANS32 primers.

Helgason et al. [49] designed the general fungal primer AM1 (Table 4) which could be used to detect AMF actually colonizing plant roots. This is because AM1 targets the 18S rDNA of AMF, and exclude plant DNA sequences. To date, data on the genetic variation of the AMF using the AM1 primer is likely the largest data set available on the genetic diversity of AMF collected from diverse natural environments. The AM1 has been shown to amplify three families of the AMF (Glomeraceae, Gigasporaceae and Acaulosporaceae). Most studies of AMF communities and host specificity used the combination of the universal and AMF specific NS31-AM1 primers to amplify the central region of the 18S rDNA [18, 20]. However, it was reported by Redecker et al. [27] and Schüßler et al. [22] that the primer AM1 does not fit all AM fungal taxa, and that AM1 is not specific to all AMF. They mentioned that the AM1 primer is specific to the AM fungi of orders Glomerales and Diversisporales, but not Archaeosporales and Paraglomerales. Daniell et al. [28] mentioned that the NS31-AM1 primer pair could amplify most taxa of the Glomeromycota, but also exclude the basal families Archaeosporaceae and Paraglomaceae.

Redecker [50] designed some specific PCR primers (Tables 4 and 5) to identify divergent clades of AMF within colonized roots. The primers targeted at five major phylogenetic subgroups of AMF (*Glomus*, *Acaulospora*, *Entrophospora*, *Scutellospora* and *Sclerocystis*). They also could facilitate specific amplification of ITS and 18S rDNA fragments from colonized roots in the absence of spores. Members of these groups were identified after analysis of RFLP patterns. In a phylogenetic study, Redecker et al. [27] designed primers LOCT670R and ATRP420, which allowed to detect some *Paraglomus* and *Glomus* species and spores belonging to *Archaeospora*, *Acaulospora* and *Zygomicetes*, respectively.

Saito et al. [51] designed primers to successfully amplify Glomeromycota, both *Miscanthus* and *Zoysia*, roots. They used the primers in a nested PCR; the

Table 4. Specific primers used to amplify fragments of the 18S rDNA region of arbuscular mycorrhizal fungi

Primer name	Primer sequence	Target group	rDNA region	Reference
VALETC	5'-ATCACCAAGGTTTAGTTGGTTGC-3'	Taxon-specific (G. etunicatum)	18S	[47]
VAGIGA	5'-TCACCAAGGGAAACCCGAAGG-3'	Family-specific (Gigasporaceae)	18S	[47]
VAGLO	5'-CAAGGGAATCGGTTGCCCGAT-3'	Taxon-specific (Glomus sp.)	18S	[47]
VAACAU	5'-TGATTCACCAATGGGAAACCCC	Family-specific (Acaulosporaceae)	18S	[47]
VANS22	5'-TAAACACTCTAATTTTTCAA	Eukaryotes, fungi and AMF	18S	[47]
VANS32	5'-AAGCTCGTAGTTGAATTTCGG-3'	Eukaryotes, fungi and AMF	18S	[47]
VANS1	5'-GTCTAGTATAATCGTTATACAGG-3'	AMF taxon-specific (Glomus and Gigaspora)	18S	[45]
AM1	5'-GTTTCCCGTAAGGCGCCGAA-3'	General fungal and AMF: Glomeraceae, Gigasporaceae and Acaulosporaceae	18S	[49]
GLOM1310	5'-AGCTAGGYCTAACATTGTTA-3'	G. mosseae, G. intrarradices	Middle of 18S	[50]
LETC1670	5'-GATCGGCGATCGGTGAGT-3'	G. etunicatum, G. claroideum	Final of 18S	[50]
ACAU1660	5'-TGAGACTCTCGGATCGG-3'	Acaulosporacea sensu stricto	Final of 18S	[50]
ARCH1311	5'-TGCTAAAATAGCCAGGCTGY-3'	A. gerdemannii/A. trappei group	Middle of 18S	[50]
		G. occultum/G. brasilianum group		
LOCT670R	5'-AAGGCCATGACGCTTCGC-3'	Paraglomus and Glomus	18S	[27]
ATRP420	5'-AACAATACAGGGCCTTTAC-3'	Archaeospora, Acaulospora and Zygomicetes	18S	[27]
GLO1375R	5'-ACTTCCATCGGTTAAACACC-3'	Glomus	Middle of 18S	[52]
ARCH1375R	5'-TCAAACTTCCGTTGGCTARTCGCRC-3'	Archaeospora	Middle of 18S	[52]
ML1	5'-AACTTTCGATGTGTAGGATAGA-3'	Fungi-specific	18S	[67]
AML2	5'-CCAAACACTTTGGTTTCC-3'	Fungi-specific	18S	[67]
AMV4.5F	5'-AATTGGAGGGCAAGTCTGG-3'	Eukaryota and AMF	First middle of 18S (V3)	[51]
AMV4.5R	5'-AGCAGGTTAAGGTCTCGTTCGT-3'	Eukaryota and AMF	End middle of 18S (V7)	[51]
AMV4.5NF	5'-AAGCTCGTAGTTGAATTTCG-3'	Zygomycota and AMF	First middle of 18S (V4)	[51]
AMV4.5NR	5'-CACCCATAGAATCAAGAAAGA-3'	Zygomycota and AMF	End middle of 18S (V7)	[51]
AMDGR	5'-CCCAACTATCCCTATTAATCAT-3'	AMF (Archaeospora)	18S (used for DGGE)	[68]
FM6	5'-ACCTGCTAAATAGTCAGGCTA-3'	Gigasporaceae	18S (near to V9)	[35]

(Contd.)

Primer name	Primer sequence	Target group	rDNA region	Reference
MNS1	5'-TGCATGTCTAAGTATAAACCATTTATACAGG-3'	*Glomus*	End middle of 18S	[53]
MNS4	5'-TCCCTAGTCGGCATAGTTTATGGT-3'	*Glomus*	End middle of 18S	[53]
GLOMBS1670	5'-AGCTTTAAACCGGCATCTGT-3'	*G. mosseae* subgroup within *Glomus* group A	18S	[54]
PARA1313	5'-CTAAATAGCCAGGCTGTTCTC-3'	*Paraglomus*	18S	[54]
GlomerWT0	5'-GDWTCATTCAAATTTCTGCCCTAT-3'	AMF	18S	[31]
Glomer1536	5'-RTTGCAATGCTCTATCCCCA-3'	AMF	18S	[31]
GlomerWT3	5'-CAAACTTCCATTGRCTAAATGCCA-3'	Diversisporaceae	18S	[31]
GlomerWT4	5'-CAAACTTCCATBGGCTAAACGCCR-3'	Glomeraceae, Gigasporaceae, Pacisporaceae, Paraglomeraceae	18S	[31]
GlomerWT1	5'-CAAACTTCMGTTGGCTAATCGCGC-3'	Archaeosporaceae	18S (Arch1375 modified)	[31]
GlomerWT2	5'-CAAACTTCCATCGGTTARACACCG-3'	Glomeraceae, Pacisporaceae	18 S (Arch1375 modified)	[31]

Table 5. Specific primers used to amplify fragments of the Internal transcribed spacer (ITS) region of arbuscular mycorrhizal fungi

Primer name	Primer sequence	Target group	rDNA region	Reference
GMOS1	5'-CTGANGACGCCAGGTCAAAC-3'	*G. mosseae, G. monosporum*	ITS	[55]
GMOS2	5'-AAATATTTAAAACCCCACTC-3'	*G. mosseae, G. monosporum*	ITS	[55]
GMOS3	5'-CGACGCGATCACCCTNAAAAA-3'	*G. mosseae, G. monosporum*	ITS	[55]
GMOS4	5'-GCGAGGCTTGCGAAAATA-3'	*G. mosseae, G. monosporum*	ITS	[55]
GLOM5.8R	5'-TCCGTTGTTGAAAGTGATC-3'	*G. mosseae, G. intraradices*	ITS (5.8)	[50]
GIGA5.8R	5'-ACTGACCCTCAAGCAKGTG-3'	Gigasporacea	ITS (5.8)	[50]
GOCC56	5'-CAACCCGCTCKTGTATTT-3'	*G. occultum, G. brasilianum*	ITS and 5.8S	[56]
GOCC427	5'-CCACACCCAKTGCGC-3'	*G. occultum, G. brasilianum*	ITS and 5.8S	[56]
GBRAS-86	5'-TGTATTGGATCAAACGTC-3'	*G. brasilianum, Glomus* (one strain)	ITS and 5.8S	[56]
GBRAS-388	5'-CGCTATTCATTGTGCACT-3'	*G. brasilianum, Glomus* (one strain)	ITS and 5.8S	[56]
SSU-Glom1	5'-ATTACGTCCCTGCCCTTTGTACA-3'	Glomeromycota (except Archaeosporaceae)	ITS	[25]
LSU-Glom1	5'-CTTCAATCGTTTCCCTTTCA-3'	Glomeromycota (except Archaeosporaceae)	ITS	[25]
MarGR1	5'-ACGTTCGAAAAATCATGCAAAATT-3'	*Glomus*	ITS1	[53]
LSU-Glom1b	5'-TCGTTTCCCTTTCAACAATTTCAC-3'	Archaeosporales (Glomeromycota): *Ambispora fennica*	ITS	[69]

outer primer pair in the first reaction was AMV4.5F and AMV4.5R, and the inner primer pair in the second reaction was AMV4.5NF and AMV4.5NR. Nested PCR amplification products (about 650 bp) were obtained from fungal DNA. However, like the AM1 primer, primers are not specific for all AM fungal species.

Russell et al. [52] designed primers (Table 4) which aimed to find species of Glomales present in the mycorrhizal root nodules of four species of Podocarpaceae (New Zealand rain forest). Primers targeted in the middle of 18S rDNA of several AMF belonging to the newly characterised family Archaeosporaceae, and two lineages of Archaeospora. Later, Russell and Bulman [53] designed PCR primers (AMF specific) which matched within the ITS spacer sequences of the *Glomus* phylotypes in symbiosis with *Marchantia foliácea*. Hijri et al. [54] designed the primers GLOMBS1670 and PARA1313 which successfully detect the *G. mosseae* subgroup within the *Glomus* group A and the genus *Paraglomus*, respectively. They used a nested PCR from environmental samples of arable soils. Wubet et al. [31] compared the diversity of AMF associated with *Juniperus procera* from two geographically separated sites in the dry Afromontane forests of Ethiopia based on the analysis of the 18S rDNA. Firstly they used the NS31–AM1 primer pair. They designed a nested PCR approach with a series of newly designed and modified specific primers to amplify approximately 1130 bp of the 18S rDNA of the AMF colonizing *J. procera*. The first amplification of fungal DNA was performed using the primer pair GlomerWT0 and Glomer1536. GlomerWT0 and Glomer1536 match with a wide range of higher fungi. For the second reaction they used the forward primer GlomerWT0 in combination with specific reverse primers GlomerWT1, GlomerWT2, GlomerWT3 and GlomerWT4. This successfully amplified part of the 18S rDNA of some AMF belonging to Diversisporaceae, Glomeraceae, Gigasporaceae, Pacisporaceae, Paraglomeraceae and Archaeosporaceae.

Millner et al. [55, 56] designed primers (Table 5) which successfully targeted the 5.8 S subunit and flanking ITS regions, and 18S rDNA from spores of some *G. mosseae* and *G. monosporum* isolates, and the two ancient AMF *G. occultum* and *G. brasilianum* from highly diluted extracts of colonized roots. However, they could not amplify most of tested *Glomus* species. Van Tuinen et al. [23, 57] designed general (Table 3) and AMF specific primers for the *G. mosseae, G. intraradices, S. castanea,* and *G. rosea* groups (Table 6), which targeted the D2 domain of the 25S rDNA. Trouvelot et al. [58] also designed general and fungus-specific primers to detect rDNA of *G. mosseae* and *G. rosea*; they were visualized using digoxigenin-labeled 25S rDNA probes obtained by nested PCR. Gamper and Leuchtman [59] designed two taxon-specific primer pairs (Table 6) to specifically detect *A. paulinae* (f6/r1) and a currently undescribed AMF taxon of *Glomus* sp. (f4/r2) (member of the Diversisporaceae); these were used in a nested PCR procedure. The nested PCR amplification comprised two steps: (1) amplification with the universal

Table 6. Specific primers used to amplify fragments of the 25S rDNA region of arbuscular mycorrhizal fungi

Primer name	Primer sequence	Target group	rDNA region	Reference
5.21	5'-CCTTTTGAGCTCGGTCTCGTG-3'	G. mosseae	D2 domain of the 25S rDNA	[23]
8.22	5'-AACTCCTCACGCTCCACAGA-3'	G. intraradices	D2 domain of the 25S rDNA	[57]
4.24	5'-TGTCCATAACCAACTTCGT-3'	S. castanea	D2 domain of the 25S rDNA	[57]
23.22	5'-GAATCACAGTCAGCATGCTA-3'	G. rosea	D2 domain of the 25S rDNA	[23]
5.23	5'-GTACGGTTAGTCAACATCG-3'	G. mosseae and some Glomus sp.	25 S rDNA	[23]
5.25	5'-ATCAACCTTTTGAGCTCG-3'	G. mosseae	25 S rDNA	[58]
23.46	5'-GCTATCCGTAATCCAATACTG-3'	G. rosea	25 S rDNA	[58]
LSU RK4	5'-GGGAGGTAAATTTCTCCTAAGGC-3'	G. mosseae	D2 domain of the 25S rDNA	[29]
LSU3f	5'-AGTTGTTTGGGATTGCAGC-3'	Glomus (some sp.)	D2 domain of the 25S rDNA	[29]
LSU4f	5'-GGGAGGTAAATTTCTCCTAAGGC-3'	Glomus (some sp.)	D2 domain of the 25S rDNA	[29]
LSU6f	5'-AAATTGTTGAAAGGGAAACG-3'	Glomus (some sp.)	D2 domain of the 25S rDNA	[29]
LSU9f	5'-ATTCGTTAAGGATGTTGACG-3'	Glomus (some sp.)	D2 domain of the 25S rDNA	[29]
LSU5r	5'-CCCTTTCAACAATTTCACG-3'	Glomus (some sp.)	D2 domain of the 25S rDNA	[29]
LSU7r	5'-ATCGAAGCTACATTCCTCC-3'	Glomus (some sp.)	D2 domain of the 25S rDNA	[29]
LSU8r	5'-GGGTATCCGTTGCAATCCTC-3'	Glomus (some sp.)	D2 domain of the 25S rDNA	[29]
LSU 0805	5'-CATAGTTCACCATCTTTCGG-3'	Glomus (some sp.)	5'end of the 25 S	[29]
ALF01	5'-GGAAAGATGAAAAGAAACTTTGAAAAGAG-3'	G. coronatum	D2 domain of the 25S rDNA	[70]
38.21	5'-TGGGCTCGCGGCCGGTAG-3'	G. claroideum	25 S rDNA	[30]
cad 4.1	5'-TCGAGTATTGCTGCGACGA-3'	Glomus sp. (near to Glomus gerdemannii Rose)	25 S rDNA	[30]
cad 4.2	5'-CTCAAGTGTCCACAACTGC-3'	Glomus sp. (near to G. gerdemannii)	25 S rDNA	[30]
cad 5.1	5'-GAAGTCTGTCGCAGTCTG-3'	Glomus sp. (near to G. occultum)	25 S rDNA	[30]

(Contd.)

Table 6. (*Contd.*)

Primer name	Primer sequence	Target group	rDNA region	Reference
cad 5.3	5′-TCG-CGA-AAG-CTTGTG-3′	*Glomus* sp. (near to *G. occultum*)	25 S rDNA	[30]
FLR3	5′-TTGAAAGGGAAACGATTGAAGT-3′	*Glomus* (groups A and B), Gigasporaceae and Acaulosporaceae (not Archaeospora)	25S rDNA	[71]
FLR4	5′-TACGTCAACATCCTTAACGAA-3′	*Glomus* (groups A and B), Gigasporaceae and Acaulosporaceae (not Archaeospora)	25S rDNA	[71]
f6	5′-TAAATCTCCGAGGTTTCCTTGGC-3′	*A. paulinae*	5′ end of the 25S rDNA	[59]
r1	5′-TCATCTTTCCCTCACGGTACTTG-3′	*A. paulinae*	Near to domain D1 of the 25S	[59]
f4	5′-TAAATCTACCTGGTTCCCAGGTC-3′	*Glomus* sp. (member of the Diversisporaceae)	5′ end of the 25S rDNA	[59]
r2	5′-TGAACCCAAAACCACCAAACTG-3′	*Glomus* sp. (member of the Diversisporaceae)	Near to D2 domain of 25S	[59]

Table 7. Specific primers used for minisatellite analysis of arbuscular mycorrhizal fungi

Primer name	Primer sequence	Target group	Reference
M13	5′-GAGGGTGGCGGTTCT-3′	Minisatellite of *G. margarita*	[60]
AM1-2	5′-GTT TCC CGT AAG CGC CGA A-3′	Minisatellite of *Gigaspora* sp.	[61]

fungal primer pair LR1/FLR2 and (2) amplification with the newly designed specific primer pairs. The primers targeted the 5' end of the 25S rDNA, where they flank the variable domain D1.

Primers for DGGE Analysis

De Souza et al. [35] used the DGGE to assess *Gigaspora* diversity, and distribution and competitiveness of *Gigaspora* spp. by screening 48 isolates from culture and soil field samples. The study revealed differences in the genomic variation of the V region of the 18S rDNA gene. They designed a new primer (Table 4), and found that the V9 region could be used for reliable identification of all recognized species within this genus. Sato et al. designed a new primer for PCR-DGGE analysis from published 18S rDNA sequences of mycorrhiza. This allowed discrimination of five species of AMF (*G. claroideum*, *G. clarum*, *G. etunicatum*, *G. margarita* and *Archaeospora leptotichaa*) from spores collected in a grassland. The primer pair (GC-AMV4.5NF/AMDGR) allowed to amplify approximately 300 bp fragments corresponding to part of the 18S rDNA gene of *Archaeospora leptoticha*, *G. claroideum*, *G. etunicatum*, *G. clarum* and *G. margarita*. Schwarzott and Schussler [46] designed the general fungal primers GeoA2 and Geo11 (Table 1) to amplify an approximately 1.8 kb fragment of the 18S rDNA gene. Ma et al. [36] successfully used the primers of Schwarzott and Schussler [46] by a nested PCR approach for DGGE analysis of DNA isolated from spore and soil samples. They used GeoA2 and Geo11 primers for the first PCR reaction, and the primers AM1 and NS31-GC for the second PCR reaction. This produced an approximately 550 bp fragment.

Primers for Minisatellite Analysis

Zézé et al. [60] used the M13 minisatellite-primer (Table 7) for fingerprint analyses to detect the presence of spores of *Gigaspora* in a mixed population. Yokoyama et al. [61] designed an oligonucleotide probe based on the DNA sequence of *G. margarita* to investigate the auto-ecology of a strain of a commercial inoculum. They designed the primer AM1-2 after modifying AM1 to match better the DNA sequence of *Gigaspora* sp. They tested the success of the primer using the obtained amplicon for mini-satellite analysis with the M13 primer.

Conclusion

In the field, AMF mycelium is embedded deeply within plant roots and other soil microorganisms. Therefore, DNA extraction is a problem; as a result, numerous pathogenic and saprophytic fungi will be co-detected. The limitations of the original primers restricted their application in molecular analysis of AMF communities, and resulted in the design and development of several new primers. Ribosomal DNA primers have become a widely-employed technique for detecting various organisms present in low amounts in complex samples. For over twenty years, researchers have designed a range of rDNA primers specific for the detection of the AMF with meticulous work. Several of the universal or general fungi primers were developed to amplify a broad taxonomic range. However, most of them were preferentially designed to amplify specific groups of fungi such as AMF. Designing one primer for all glomalean fungi excluding plants and other fungi proved to be difficult. Obtaining results not always should be expected. It must be emphasized that no single set of primers or community profiling technique will be optimal to access fungal diversity in all instances. Actually, there are many specific primers which help identification of AMF from environmental samples. However, many groups have not yet been identified which is a subject for further research.

References

1. Mosse, B., Stribley, D.P. and Le Tacon, F. *Adv Microb Ecol* 1981, 5:137-210.
2. Van der Heijden, M.G.A., Wiemken, A. and Sanders, I.R. *New Phytol* 2003, 157:569-578.
3. Van der Heijden, M.G.A., Boller, T., Wiemken, A. et al. *Ecology* 1998, 79:2082-2091.
4. Moora, M., Öpik, M., Sen, R. et al. *Funct Ecol* 2004, 18:554-562.
5. Morton, J.B. and Benny, G.L. *Mycotaxon* 1990, 37:471-491.
6. Merryweather, J.W. and Fitter, A.H. *New Phytol* 1998, 138:131-142.
7. Thorn, R.G., Reddy, C.A., Harris, D. et al. *Appl Environ Microbiol* 1996, 62:4288-4292.
8. Smith, S.E. and Read, D.J. Mycorrhizal Symbiosis. Academic Press, Cambridge, 1997.
9. Clapp, J.P., Young, J.P.W., Merryweather, J.W. et al. *New Phytol* 1995, 130:259-265.
10. Rosendahl, S. and Stukenbrock, E.H. *Mol Ecol* 2004, 13:3179-3186.
11. Mullis, K.B. and Faloona, F.A. *Method Enzymol* 1987, 155:335-350.
12. Lanfranco, L., Perotto, S. and Bonfante, P. *In:* Applications of PCR in Mycology. Bridge, P.D., Arora, D.K., Reddy, C.A. et al. (eds), CAB International, UK, 1998, pp. 107-124.
13. Mahuku, G.S., Platt, H.W. and Maxwell, P. *Can J Plant Pathol* 1999, 21:125-131.
14. Smit, E., Leeflang, P., Glandorf, B. et al. *Appl Environ Microbiol* 1999, 65:2614-2621.
15. van Elsas, J.D., Duarte, G.F., Keijzer-Wolters, A. et al. *J Microbiol Meth* 2000, 43:133-151.
16. Morton, J.B. and Redecker, D. *Mycologia* 2001, 93:181-195.
17. Sanders, I.R., Alt, M., Groppe, K. et al. *New Phytol* 1995, 130:419-427.
18. Helgason, T., Fitter, A.H. and Young, J.P.W. *Mol Ecol* 1999, 8:659-666.
19. Kuhn, G., Hijri, M. and Sanders, I.R. *Nature* 2001, 414:745-748.

20. Vandenkoornhuyse, P., Baldauf, S.L., Leyval, C. et al. *Science* 2002, 295:2051.
21. Jansa, J., Mozafar, A., Banke, S. et al. *Mycol Res* 2002, 6:670-681.
22. Schüßler, A., Gehrig, H., Schwarzott, D. et al. *Mycol Res* 2001, 105:5-15.
23. Van Tuinen, D., Jacquot, E., Zhao, B. et al. *Mol Ecol* 1998, 7:879-887.
24. Redecker, D. *Plant Soil* 2002, 244:67-73.
25. Renker, C., Heinrichs, J., Kaldorf, M. et al. *Mycorrhiza* 2003, 13:191-198.
26. Simon, L., Lalonde, M. and Bruns, T.D. *Appl Environ Microbiol* 1992, 58:291-295.
27. Redecker, D., Morton, J.B. and Bruns, T.D. *Mol Phyl Evol* 2000, 14:276-284.
28. Daniell, T.J., Husband, R., Fitter, A.H. et al. *FEMS Microbiol Ecol* 2001, 36:203-209.
29. Kjøller, R. and Rosendahl, S. *Plant Soil* 2000, 226:189-196.
30. Turnau, K., Ryszka, P., Gianinazzi-Pearson, V. et al. *Mycorrhiza* 2001, 10:169-174.
31. Wubet, T., Wei, M., Kottke, I. et al. *Mycol Res* 2006, 110:1059-1069.
32. Vandenkoornhuyse, P., Ridgway, K.P., Watson, I.J. et al. *Mol Ecol* 2003, 12:3085-3095.
33. Kowalchuk, G.A., Gerards, S. and Woldendorp, J.W. *Appl Environ Microbiol* 1997, 63:3858-3865.
34. Kowalchuk, G.A., de Souza, F.A. and van Veen, J.Á. *Mol Ecol* 2002, 11:571-581.
35. de Souza, F.A., Kowalchuk, G.A., Leeflang, P. et al. *Appl Environ Microbiol* 2004, 70:1413-1424.
36. Ma, W.K., Siciliano, S.D. and Germida, J.J. *Soil Biol Biochem* 2005, 37:1589-1597.
37. Mummey, D.L., Rillig, M.C. and Holben, W.E. *Plant Soil* 2005, 271:83-90.
38. Mummey, D.L. and Rillig, M.C. *J Microbiol Meth* 2007, 70:200-204.
39. Muyzer, G., de Waal, E.C. and Uitterlinden, A.G. *Appl Environ Microbiol* 1993, 59:695-700.
40. Vainio, E.J. and Hantula, J. *Mycol Res* 2000, 104:927-936.
41. Zézé, A., Dulieu, H. and Gianinazzi-Pearson, V. *Mycorrhiza* 1994, 4:251-254.
42. Zézé, A., Hosny, M., Gianinazzi-Pearson, V. et al. *Appl Environ Microbiol* 1996, 62:2443-2448.
43. White, T.J., Bruns, T., Lee, S. et al. *In:* PCR protocols, a guide to methods and applications. Innis, M.A., Gelfand, D.H., Sminsk, J.J. et al. (eds.), Academic Press, San Diego, California. 1990, 315-322.
44. Bousquet, J., Simon, L. and Lalonde, M. *Can J Forest Res* 1990, 20:254-257.
45. Simon, L. and Lalonde, M. DNA probes for the detection of arbuscular endomycorrhizal fungi. United States Patent 5434048. 1995, http://www.freepatentsonline.com /5434048.html
46. Schwarzott, D. and Schussler, A. *Mycorrhiza* 2001, 10:203-207.
47. Simon, L., Lévesque, R.C. and Malonde, M. *Appl Environ Microbiol* 1993, 58:4211-4215.
48. Simon, L. *New Phytol* 1996, 133:95-101.
49. Helgason, T., Daniell, T.J., Husband, R. et al. *Nature* 1998, 384:431.
50. Redecker, D. *Mycorrhiza* 2000, 10:73-80.
51. Saito, K., Suyama, Y., Sato, S. et al. *Mycorrhiza* 2004, 14:363-373.
52. Russell, A.J., Bidartondo, M.I. and Butterfield, B.G. *New Phytol* 2002, 156:283-295.

53. Russell, A.J. and Bulman, S. *New Phytol* 2005, 165:567-579.
54. Hijri, I., Korova, Z., Oehl, F. et al. *Mol Ecol* 2006, 15:2277-2289.
55. Millner, P.D., Mulbry, W.W., Reynolds, S.L. et al. *Mycorrhiza* 1998, 8:19-27.
56. Millner, P.D., Mulbry, W.W. and Reynolds, S.L. *FEMS Microbiol Lett* 2001, 196:165-170
57. van Tuinen, D., Zhao, B. and Gianinazzi-Pearson, V. *In:* Mycorrhiza manual. Varma, A. (ed.), Springer, Heidelberg, 1998, 387-399.
58. Trouvelot, S., van Tuinen, D., Hijri, M. et al. *Mycorrhiza* 1999, 8:203-206.
59. Gamper, H. and Leuchtmann, A. *Mycorrhiza* 2007, 17:145-152.
60. Zézé, A., Sulistyowati, E., Ophel-keller, K. et al. *Appl Environ Microbiol* 1997, 63:676-678.
61. Yokoyama, K., Tateishi, T., Marumoto, T. et al. *FEMS Microbiol Lett* 2002, 212:171-175.
62. Gargas, A. and Taylor, J.W. *Mycologia* 1992, 84:589-592.
63. Gardes, M. and Bruns, T.D. *Mol Ecol* 1993, 2:113-118.
64. Vilgalys lab, Duke University. Conserved primer sequences for PCR amplification and sequencing from nuclear ribosomal RNA. http://www.biology.duke.edu/fungi/mycolab/primers.htm
65. Howlett, B.J., Brownlee, A.G., Guest, D.I. et al. *Curr Genet* 1992, 22:455-461.
66. Gargas, A. and DePriest, P.T. *Mycologia* 1996, 88:745-748.
67. Lee, J.K. MS Dissertation. Korea National University of Education, Cheongweon, Korea, 2003.
68. Sato, K., Suyama, Y., Saito, M. et al. *Grassland Sci* 2005, 51:179-181.
69. Walker, C., Vestberg, M., Demircik, F. et al. *Mycol Res* 2007, 111:137-153.
70. Clapp, J.P., Rodriguez, A. and Dodd, J.C. *New Phytol* 2001, 149:539-554.
71. Gollotte, A., van Tuinen, D. and Atkinson, D. *Mycorrhiza* 2004, 14:111-117.

14

The Impact of Climate Changes on Belowground: How the CO$_2$ Increment Affects Arbuscular Mycorrhiza

Fernanda Covacevich and Ricardo Luis Louro Berbara
Universidade Federal Rural do Rio de Janeiro, Dep. de Solos, BR 465, km 7
CEP 23890-970 Seropédica, RJ, Brasil; Inter-American Institute for Global
Change Research (IAI) CRN II/14 - Supported by the US National Science
Foundation (Grant GEO-04523250)

Introduction

The world is changing and during the last years people are speeding up the process. Studies on how climate is changing, and how these changes differ along latitudes, are much more advanced than studies on how global change will affect plant community structure and how it relates to soil symbionts along different ecosystems. This is because changes that occur aboveground are easier to measure than those occurring belowground. Arbuscular mycorrhizal fungi (AMF) are belowground plant symbionts; they are a powerful link in the chain of transfers by which carbon (C) moves from the atmosphere to the plant and finally to soil sinks [1, 2]. They can potentially change C cycling rates by influencing plant growth and plant diversity [3]. Thus, any factor that changes mycorrhizal functions undoubtedly will affect primary productivity and soil C stock. Although it is widely acknowledged that AMF have multiple effects on terrestrial ecosystems, their relative contribution to ecosystem processes is unknown. As environmental conditions will change globally in the next few decades, probably at unprecedent rates, the behaviour of those AMF will play a substantial part in the response of ecosystems. Furthermore, because AMF represent an interface between the soil-plant system, their potential ability to regulate plant response to global change is one key reason why their responses need to be understood. We need to identify major obstacles that prevent us to gain a full understanding where our knowledge is surprisingly poor. There is an obvious need to understand the impact that global change, especially increases in atmospheric carbon dioxide (CO_2), will have on mycorrhizal symbioses. In this chapter we will focus on recent advances in our understanding of mycorrhizal association responses to environmental change, mainly to increases in atmospheric CO_2.

Nutrient, High Temperature and CO_2 Effects on Plant Growth and Mycorrhiza

Atmospheric CO_2 concentrations, nitrogen (N) deposition and global average temperature are increasing over the last century as a result of anthropogenic emissions of greenhouse gases and changes in land use [4]. The amount of C stored globally in soils is much larger than that in vegetation, soil being the major organic C source. Subsequently, changes in atmospheric CO_2 undoubtedly will produce changes on soil organisms, nutrients and organic matter dynamics. The atmospheric CO_2 increment has led to increase in global ecosystem C storage [5]. This increase affects terrestrial ecosystems because they are a network intimately connected to atmospheric CO_2 levels through photosynthetic fixation of CO_2, sequestration of C into biomass and soil, and subsequent release of CO_2 through respiration and organic matter decomposition. Translocation of C to the rhizosphere via AMF will also influence food web interactions which subsequently will mediate ecosystem feedbacks that regulate cycling of C and N.

Abiotic factors such as soil fertility [6, 7] and temperature [8] can strongly impact plant and arbuscular mycorrhiza (AM) functioning. The phosphorus (P) supply may also function as a secondary limit, and also limits the net primary production. It is well established that available soil P, mainly from inorganic or synthetic sources, could increase plant growth but depress AMF [7, 9-11]. However, Menge and Field [12] after analyzing published P cycle responses to elevated CO_2, N, precipitation or temperature in 16 ecosystems, ranging from deserts to rainforests, and from the tropic to the arctic, concluded that neither CO_2 nor temperature directly affect P demand or limitation. They showed that only N deposition increased plant growth, and it increased P limitation to the dominant grass and ecosystem-level P demand.

While CO_2, N, and water frequently limit plant growth [13], warming can either increase or decrease primary production. A number of studies have shown that soil temperature, moisture, and nutrient availability control, in part, the timing and duration of root growth [14-19]. Soil temperature probably has a strong role in determining root turnover because the costs of root maintenance increase exponentially with temperature [18, 20-23]. Gill and Jackson [24] concluded that both aerial plant growth and root turnover will probably be sensitive to many of the factors considered in global change analyses because they are a central component of ecosystem C and nutrient cycling [25, 26]. They found that root turnover in grasslands, shrub lands, and fine root forests were positively correlated with mean annual temperature in many terrestrial ecosystems across broad climatic gradients. The strength of some of those relationships might be the result of a strong correlation between air and soil temperatures in systems with low leaf area (e.g., grasslands and shrublands).

Current information suggests that responses to elevated atmospheric CO_2 will be largely controlled by host-plant responses, but that AMF will respond directly to elevated soil temperature. Although our knowledge of AM fungal response to temperature is slight, it was reported that increases in soil temperature can increase the AMF extraradical mycelium. The direct response of AMF to temperature may have large implications for rates of C cycling. New evidence shows that AM fungal hyphae may be very short lived, potentially acting as a rapid route by which C may cycle back to the atmosphere. Gavito et al. [8] found that the extra radical mycelium of *Glomus caledonium* did not grow at all at 10°C, but grew well at 25°C. Heinemeyer and Fitter [27] used a compartmentalized experimental design (in which only part of the soil volume was accessible to the fungus *Glomus mosseae* and *G. hoi* because of a dividing mesh) to measure the effects of soil temperature on the fungus alone. In this case, growth of the extraradical mycelium of both AMF species was stimulated by warming, but root colonization of *P. lanceolata* (unwarmed plant) was unaffected. Growth of the fungus appeared to be strongly regulated by the host when these authors studied the effects of CO_2 availability. Several studies have shown an increased percentage of AM colonization at elevated CO_2 atmospheric concentrations [28, 29]. Temperature, however, did not alter the length of extraradical mycelium per unit length of colonized root [27]. Staddon et al. [30] found that warming increased root colonization but decreased the length of extraradical mycelium. They increased field soil temperature only by 3°C during the winter months in temperate grasslands (northern England) by using soil-warming cables: increases in mycorrhizal colonization because of higher temperatures were probably due to vegetational changes rather to direct temperature effects on AMF. Fitter et al. [31] mentioned that aspects of climate change other than CO_2 might have important direct effects on AMF. If soils warm up, fungi growth might be affected; for example, some fungi might become active at times of the year when they are often dormant.

Results are contradictory and do not allow general conclusions. One impact of warming is to reduce soil moisture. At the same time, warming probably results in increased N mineralization and hence greater root growth. Changes in soil moisture content, either drier or wetter conditions, would almost certainly have large, but yet unpredictable, effects on AMF. In a warmer climate, for example, plant C fixation might increase, with the same impacts as elevated CO_2. Alternatively, C fixation or fungal growth might be depressed, or there might be qualitative effects on fungal community structure or on its temporal pattern (phenology). However, if increases in net primary productivity persist over time (as a result of increases in CO_2, temperature, nutrients and others), soil C inputs will ultimately increase via enhanced plant biomass turnover and/or root C losses to the soil. Organic C supply might then limit soil microbial activity, and relevant changes in soil C cycling could therefore be expected.

Are AMF Involved in Global Change?

Scientists have predicted that climate warming could increase heterotrophic respiration, resulting in a greater CO_2 efflux to the atmosphere [32]. However, not all microbes are consumers of organic matter. The mycorrhizal associations are widespread plant root symbioses that depend upon the C which is fixed by photosynthesis and then translocated from the plant to the fungi. Then, AMF do not break down soil organic matter, and should be treated as a special case in global change.

It was reported that (1) plant genotypes differ in mycorrhizal dependency [33]; (2) mycorrhizal fungal genotypes vary in mutualistic effects [34]; and (3) host plants, AMF and environmental conditions interact to control the costs and benefits of mycorrhizal symbioses [35]. Field studies suggest that changes in vegetation in response to environmental change may play the largest role in determining the structure of the AM fungal community. Photosynthesis is more limited by the availability of atmospheric CO_2 concentrations in C_3 than C_4 plants [36]. Fast-growing, C_3 plant species benefit more from elevated CO_2 than from mycorrhizae than slow-growing plants with a C_4 photosynthetic pathway [37-40]. Johnson et al. [40] reported that elevated CO_2 increased species richness of the plant community (of a mesocosm study) when AMF were present, but not when they were absent. The survival of slow-growing mycotrophic forbs was increased at elevated CO_2 levels, suggesting that the increased CO_2 ameliorated the C cost of the symbiosis. However, our understanding of soil-borne microbial communities remains rather poor. This is because of their extraordinary complexity [41] and difficulties associated with laboratory cultivation of most soil microbes. Taxa of AMF can also differ in their responses to CO_2 enrichment [42-44].

Several studies support the hypothesis that atmospheric CO_2 enrichment will influence the balance between mycorrhizal costs and benefits. From a plant perspective, AMF functioning is determined by the balance between photosynthate costs and nutrient benefits [45]. Plants grown at elevated atmospheric CO_2 typically have more soluble carbon in root pools; this is the place where the fungi obtain their carbon. Fungal responses can simply be explained as a function of plant growth: larger plants support larger fungal biomass and activity [23]. Mutualistic effects of mycorrhizal associations might be sensitive to anthropogenic enrichment of atmospheric CO_2, because it should simultaneously increase plant photosynthetic rates and soil nutrient requirements [46]. Photosynthate allocation to AMF can represent a major C cost to plants [47], and increased photosynthetic rates at elevated CO_2 should make more C available to support mycorrhizal symbioses [48, 49]. Carbon demands of AMF increase plant photosynthetic rates [50]; C gained via C assimilation through mycorrhizal enhancement is allocated to the fungus, and not to increase host plant biomass [51, 52]. On the other hand, however, sometimes the response is not clear because there are confounding effects of increases in root biomass [23, 31]. Furthermore, CO_2 enrichment can mitigate plant growth depressions which generate when C costs of AMF outweigh nutrient uptake benefits to the plant [49]. Other authors are of the opinion that elevated CO_2 often increases allocation to mycorrhizal hyphae that occur in the soil outside plant roots [3, 53]. Staddon et al. [2] concluded that any extra C assimilation would be balanced by an increased respiratory loss of the symbiosis.

Fitter et al. [31] concluded that increased plant-C fixation would increase C availability to the fungus, and that this would increase root colonization, promote fungal ability to provide P to the plant and improve plant growth. The hypothesis of Fitter et al. [31] was that the fungus existed in an environment rich in CO_2, both inside the root and in the soil, and that air CO_2 was greater than current atmospheric CO_2. However, if elevated atmospheric CO_2 increases the supply of fixed C to roots, this might promote the growth of the fungus. A positive feedback loop can be envisaged, where plants respond to elevated CO_2 by increased C fixation and subsequent transfer of more C to their roots; consequently, AMF might grow more and capture more phosphate (or perform other functions better).

Johnson et al. [46] argued that the idea that CO_2 enrichment will decrease mycorrhizal costs and/or increase benefits for plants is clearly oversimplistic because many factors mediate AMF functioning. They conducted three experiments in a greenhouse that compared the impacts of CO_2 enrichment on mycorrhizal functioning of many different plant-fungus combinations grown under high or low nutrient availability. The experiments differed in the number of study plant species (C_3, C_4, Composites, Legumes and Non-mycorrhizal forbs), use of endemic or non-endemic AM fungal inoculums, soil media, and nutrient treatments. Responses were not uniform, and results indicated that enrichment of CO_2 did not uniformly increase mycorrhizal benefits to plants. There was variability in responses but, particularly, when the AMF community was dominated by *Glomus* species, CO_2 enrichment reduced, in general, the beneficial effects of mycorrhizae on plant biomass.

Mycorrhizal fungi can potentially influence C cycling rates. The elevated atmospheric CO_2 concentrations may also indirectly affect soil structure, for instance, by affecting the concentration of binding agents in soil as glomalin-related proteins [54-56]. These changes have also the potential for a strong feedback effect on the responses of plants and soil microbes to elevated CO_2 [54]. Mycorrhizae are expected to enhance C sequestration by translocating C away from the high respiratory activity around the roots and into the soil matrix, including aggregates. Some of the structural C that is transferred to a symbiotic AMF is used to construct the extra radical mycelium. This network of fine hyphae almost certainly has a faster turnover rate than that of either roots or shoots. This fungal C could therefore be part of a rapid pathway in the C cycle that returns C to the atmosphere. A glycoprotein is produced by AMF, glomalin, which is deposited into hyphal walls of the extramatrical mycelium, and on adjacent soil particles. Glomalin was related to increases in soil aggregate stability and can remain in the soil from years to decades [56]. Glomalin is generally accumulated in soil, where it can represent up to 3-8% of soil organic C [55, 56]. As soil aggregation may protect glomalin from microbial degradation and enter to a 'slow pool' of recalcitrant soil C, an increase in aggregation proves to be an important mechanism for increasing sequestration of C [3]. Wilson et al. [57] found that AMF abundance was a major factor in explaining most of the variability in soil aggregation. They examined the role of AMF abundance on soil C and N storage, and soil aggregate formation, from intact multispecies field plots in tall grass prairie ecosystems. This study involved long-term, large-scale field manipulations (N fertilization, burning and fungicide application). They found that burning and N-fertilization increased soil AMF hyphae, glomalin-related soil protein pools, and water-stable macro aggregates. Burning increased extramatrical mycorrhizal hyphae and the production of intramatrical fungal structures (hyphae, arbuscules, coils and vesicles) in response to N enrichment. Abundance of extramatrical mycorrhizal hyphae was highly correlated with both macro- and micro-aggregates, and soil C and N. They reported that N fertilization undoubtedly contributed to C and N storage, as well as aggregate stability. They found a close relationship between AM fungal hyphal abundance, soil structure and C storage; the close correlation between AMF hyphal abundance and soil aggregation, and C and N sequestration was positive.

Treseder et al. [58], on the other hand, pointed out that as glomalin concentrations in the soil can be correlated with AM abundance, an increase in N availability might reduce carbon storage in glomalin. Alternatively, if AM fungi are directly limited by N, then AM growth and glomalin production could increase as N availability augments. They tested the hypothesis that the abundance of AMF and ectomycorrhizal fungi, as well as glomalin, decline with N addition in boreal ecosystems; this should be because plants would allocate less C to their mycorrhizal fungi when N is more readily available. The reduction in glomalin carbon after fertilization was observed only in the youngest tested sites. Therefore, the variation in glomalin responses among sites highlights the need for caution in generalizing results from the ecosystem to the biome scale. Since little is known about glomalin, it is difficult to identify potential controls with accuracy regarding its abundance in the soil.

Warnock et al. [59] reviewed the possibility of using 'terra preta de índio' (TP) soil, the fertile Amazonian Dark Earths and soil AMF as a promising tool for promoting crop growth and carbon storage [60]. However, no published data are available on the impact of TP soils on mycorrhizal functioning. Biochar (created by pyrolysis of waste biomass) can add value to non-harvested agricultural products [61, 62], and can promote plant growth [63, 64]. If biochar is returned to agricultural lands, it can increase soil C content permanently, establishing a C sink for atmospheric CO_2. Warnock et al. [59] mentioned the possibility of using biochar/mycorrhizae interactions for C sequestration in soils to contribute to climate change mitigation. This interaction could also be used for (1) restoring disturbed ecosystems, (2) land reclamation (i.e. sites contaminated by industrial pollution and mine wastes), (3) increasing fertilizer use efficiencies (with all associated economic and environmental benefits), and (4) development of methods to attain increased crop yields from sustainable agricultural activities.

Although responses are not always clear, there is no doubt that there is a strong link between climate change, and (1) C sequestration and (2) AMF. Increased fungal growth, while promoting P uptake, also increases C demands by the fungus, which might act as a regulator on these processes. Carbon availability is important for P uptake, transport and utilization. In return, P leaf status determines photosynthetic rate. This positive feedback may be eventually constrained by other deficiencies (e.g. N or water), and is therefore only likely to be important as a control on those links if P is a limiting factor. More questions than answers still remain. In order to generalize responses to global change, it is critical to identify which factors are involved with those responses.

Conclusion

Currently, humans inadvertently impact AMF through pollution and global change: increases in (1) soil temperatures, (2) N deposition and (3) atmospheric carbon dioxide (CO_2) concentrations. It is difficult to report general responses with accuracy about the effects of increased atmospheric CO_2 (all as a part of global change) on mycorrhizal symbiosis. This is because our understanding of the basic biology of the AMF is still limited. Furthermore, changes not always involve only one factor, and it is difficult to make useful predictions about the effects of the involved factors. In general, all experimental work on AMF responses to disturbances has been made using a small number of AMF species, typically *G. mosseae*, *G. intraradices*, and *Gigaspora* species. These species were chosen because they are easy to culture and grow on a wide range of hosts under standard laboratory conditions. About 150 AMF species have been described by (1) classical taxonomy [65] or (2) ribosomal DNA sequence of either spores or mycorrhizal roots [66]. However, there are many more fungi species in natural ecosystems and very few of these have been brought into culture. It means that up to date, we know AMF responses to global change just for a small number of AMF species which inhabit natural soils. It is likely that AMF responses to disturbances vary among plant and fungal taxa as well as environmental conditions. It is quite common to find that a single vegetation patch may support 30-40 AM fungal species. If fungi differ with respect to their symbiotic behaviour, they will very likely differ in response to environmental factors. A number of factors might underlie the responses of mycorrhiza to global change, particularly the increased atmospheric CO_2 concentrations. Responses can be different under different soil C:N ratios and different soil P availability levels. Climatic change may alter C allocation and turnover patterns in AMF. All this wide range of possibilities leads to the complexity of potential outcomes to global change that can be expected. A better understanding of the mechanisms underlying plant responses to increasing atmospheric CO_2 concentrations would improve our ability to predict ecosystem feedbacks to global change.

References

1. Staddon, P.L., Fitter, A.H. and Graves, J.D. *Glob Change Biol.* 1999, 5:347-358.
2. Staddon, P.L., Fitter, A.H. and Robinson, D. *J Exp Bot* 1999, 50:853-860.
3. Rillig, M.C. *Ecol Lett* 2004, 7:740-754.
4. IPCC Climate Change. Synthesis Report. Summary for Policymakers. November 2007, http://www.ipcc.ch
5. Schimel, D., Melillo, J., Tian, H. et al. *Science* 2000, 287:2004-2006.
6. Johnson, N.C. *Ecol App* 1993, 3:749-757.
7. Covacevich, F. and Echeverria, H.E. *J Plant Int* 2009, 4:101-112.
8. Gavito, M.E., Schweiger, P. and Jakobsen, I. *Glob Change Biol* 2003, 9:106-116.
9. Sinclair, A.G., Johnstone, P.D., Smith, L.C. et al. *Bioresource Technol* 2001, 79:263-271.
10. Rubio, R., Borie, F., Schalchli, C. et al. *Appl Soil Ecol* 2003, 23:245-255.
11. Covacevich, F., Marino, M.A. and Echeverria, H.E. *Eur J Soil Biol* 2006, 42:127-138.
12. Menge, D.L. and Field, C.B. *Glob Change Biol* 2007, 13:2582-2591.
13. Field, C.B., Chapin, F.S. III, Matson, P.A. et al. *Annu Rev Ecol Syst* 1992, 23:201-235.
14. Nadelhoffer, K.J., Aber, J. and Melillo, J.M. *Ecology* 1985, 66:1377-1390.
15. Vogt, K.A., Grier, C.C. and Vogt, D.J. *Adv Ecol Res* 1986, 15:303-377.
16. Pregitzer, K.S., Laskowski, M.J., Burton, A.J. et al. *Tree Physiol* 1998, 18:665-670.
17. Pregitzer, K.S., Zak, D.R., Maziasz, J. et al. *Ecol Appl* 2000, 10:18-33.
18. Hendrick, R.L. and Pregitzer, K.S. *Ecoscience* 1997, 4:99-105.
19. Nadelhoffer, K.J. *New Phytol* 2000, 147:131-139.
20. Ryan, M.G. *Ecol Appl* 1991, 1:157-167.
21. Fitter, A.H., Graves, J.D., Self, G.K. et al. *Oecologia* 1998, 114:20-30.
22. Atkin, O.K., Edwards, E.J. and Byrne, B.R. *New Phytol* 2000, 147:141-154.
23. Fitter, A.H., Heinemeyer, A., Husband, R, et al. *Can J Bot* 2004, 82:1133-1139.
24. Gill, R.A. and Jackson, R.B. *New Phytol* 2000, 147:13-31.
25. Aber, J.D., Melillo, J.M., Nadelhoffer, K.J. et al. *Oecologia* 1985, 66:317-321.
26. Bloomfield, J., Vogt, K. and Wargo, P.M. *In:* Plant roots: The hidden half. Waisel, Y., Eshel, A. and Kafkafi, U. (eds), Marcel Dekker, New York, USA, 1996, 363-381.

27. Heinemeyer, A. and Fitter, A.H. *J Exp Bot* 2004, 55:525-534.
28. Drigo, B., Kowalchuk, G.A. and van Veen, J.A. *Biol Fert Soils* 2008, 44:667-679.
29. Staddon, P.L. *New Phytol* 2005, 167:635-637.
30. Staddon, P.L., Thompson, K., Jakobsen, I, et al. *Glob Change Biol* 2003, 9:186-194.
31. Fitter AH, Heinemeyer A, Staddon PL. *New Phytol* 2000, 147:179-187.
32. Goulden, M.L., Wofsy, S.C., Harden, J.W. et al. *Science* 1998, 279:214-217.
33. Graham, J.H. and Eissenstat, D.M. *Plant Soil* 1994, 159:179-185.
34. Klironomos, J.N. *Ecology* 2003, 84:2292-2301.
35. Johnson, N.C., Graham, J.H. and Smith, F.A. *New Phytol* 1997, 135:575-585.
36. Drake, B.G., Gonzalez-Meler, M.A. and Long, S.P. *Annu Rev Plant Physiol Plant Mol Biol* 1997, 48:609-639.
37. Diaz, S. *J Biogeogr* 1995, 22:289-295.
38. Lüscher, A., Hendrey, G.R. and Nosberger, J. *Oecologia* 1998, 113:37-45.
39. Wilson, G.W.T. and Hartnett, D. *Am J Bot* 1998, 85:1732-1738.
40. Johnson, N.C., Wolf, J. and Koch, G.W. *Ecol Lett* 2003, 6:532-540.
41. Curtis, T.P., Sloan, W.T. and Scannell, J.W. *Proceedings of the National Academy of Science USA*, 2002, 99:10494-10499.
42. Klironomos, J.N., Ursic, M., Rillig, M. et al. *New Phytol* 1998, 138:599-605.
43. Treseder, K.K., Egerton-Warburton, L.M., Allen, M.F. et al. *Ecosystems* 2003, 6:786-796.
44. Wolf, J., Johnson, N.C., Rowland, D.L. et al. *New Phytol* 2003, 157:579-588.
45. Fitter, A.H. *Experientia* 1991, 47:350-355.
46. Johnson, N.C., Wolf, J., Reyes, M.A. et al. *Glob Change Biol* 2005, 11:1156-1166.
47. Koch, K.E. and Johnson, C.R. *Plant Physiol* 1984, 75:26-30.
48. Lovelock, C., Kyllo, D., Popp, M. et al. *Austr J Plant Physiol* 1997, 24:185-194.
49. Jifton, J.L., Graham, J.H., Drouillard, D.L. et al. *New Phytol* 2002, 153:133-142.
50. Wright, D.P., Scholes, J.D. and Read, D.J. *Plant Cell Environ* 1998, 21:209-216.
51. Wright, D.P., Read, D.J. and Scholes, J.D. *Plant Cell Environ* 1998, 21:881-891.
52. Miller, R.M., Miller, S.P., Jastrow, J.D. et al. *New Phytol* 2002, 155:149-162.
53. Staddon, P.L., Jakobsen, I. and Blum, H. *Glob Change Biol* 2004, 10:1678-1688.
54. Niklaus, P.A., Alphei, J., Ebersberger, D. et al. *Glob Change Biol* 2003, 9:585-600.
55. Rillig, M.C., Ramsey, P.W., Morris, S. et al. *Plant Soil* 2003, 253:293-299.
56. Rillig, M.C. and Mummey, D.L. *New Phytol* 2006, 171:41-53.
57. Wilson, G.W.T., Rice, C.W., Rillig, M.C. et al. *Ecol Lett* 2009, 12:452-461.
58. Treseder, K.K., Turner, K.E.M. and Mack, M.C. *Glob Change Biol* 2007, 13:78-88.
59. Warnock, D.D., Lehmann, J., Kuyper, T.W. et al. *Plant Soil* 2007, 300:9-20.
60. Glaser, B. *Phil Trans R Soc B* 2007, 362:187-196.
61. Major, J., Steiner, C., Ditommaso, A. et al. *Weed Biol Manag* 2005, 5:69-76.
62. Topoliantz, S., Ponge, J.-F. and Ballof, S. *Biol Fert Soils* 2005, 41:15-21.
63. Lehmann, J., Da Silva, J.P. Jr, Steiner, C. et al. *Plant Soil* 2003, 249:343-357.
64. Oguntunde, P.G., Fosu, M., Ajayi, A.E. et al. *Biol Fert Soils* 2004, 39:295-299.
65. Morton, J.B. and Benny, G.L. *Mycotaxon* 1990, 37:471-491.
66. Schüßler, A., Gehrig, H., Schwarzott, D. et al. *Mycol Res* 2001, 105:5-15.

Index

Printed and bound by CPI Group (UK) Ltd, Croydon, CR0 4YY

18/10/2024

01776243-0004